BOBSLEDS & FARMSTEADS

BOBSLEDS & FARMSTEADS

THE MEMORIES OF EVELYN GOECKE WILBERDING

BY

MERLE FRANCIS WILBERDING

PUBLISHED BY

HIGH VIEW LIBRARY
DAYTON, OHIO
1998

BOBSLEDS & FARMSTEADS

Text Copyright   ©1998   Merle Francis Wilberding
Photographs Copyright   ©1998   Merle Francis Wilberding
Illustration Copyright ©1998   Kurt Dietrick Wilberding

Library of Congress Catalog Card Number 98-93706

ISBN 0-9667588-0-3

First Edition

This edition is limited to 200 copies.

Published By

HIGH VIEW LIBRARY
33 WEST FIRST STREET
SUITE 600
DAYTON, OHIO 45402
937.449.5772
E-Mail: wilberding@coollaw.com

Printed on acid-free paper in the United States of America by
Sidney Printing Works, 2611 Colerain Avenue, Cincinnati, Ohio

CONTENTS

BIOGRAPHY

Evelyn Ludovica Goecke was born September 19, 1904, the daughter of William Goecke and Margaret Boecker Goecke. Her brothers and sisters were Frank Goecke (married Bessie Eilers), Elizabeth Goecke (married Vic Wernimont), Teresa Goecke (married Peter Oswald), Mamie Goecke (married Charles Schelle), Clarence Goecke (married Clara Murray), and Florence Goecke (married Lawrence Wolterman). She was baptized on September 25, 1904, by Father Kuemper, and her baptismal sponsors were John Walz and Ludovica Goecke.

The Goecke family lived on a farm just outside Mount Carmel, Iowa, and Evelyn Goecke graduated from the Catholic grade school in Mount Carmel. Later, the Goecke family moved to a small farm one-half mile east of Breda, Iowa, and then, in 1924, they moved to a home at 307 Bruning Street, Breda, Iowa.

On January 22, 1929, at St. Bernard's Church, in Breda, she married Anthony Francis Wilberding, the son of Frank Wilberding and Mary Wessling Wilberding. The ceremony was performed by Father J.A. Schulte, witnessed by Florence Goecke and Rudolph Wessling. Following the wedding, they lived on the Wilberding farms southwest of Breda.

Evelyn Goecke Wilberding and Anthony Francis Wilberding had ten children: Lois Wilberding (died in infancy), Alice Wilberding (married Richard "Bing" O'Brien), Evelyn Wilberding (married Lonnie Kropf), Donald A. Wilberding (married Margaret "Peggy" Waters), Edward C. Wilberding (married Maureen Murphy), Myrna C. Wilberding (married Joseph Niewohner), Marilyn F. "Sal" Wilberding (married Art Reiff), Larry J. Wilberding (married Diane Ocken), Merle Wilberding, and Shirley Wilberding (married William Jennewein).

PROLOGUE

The frigid Iowa winds had already announced the start of winter in Breda when I arrived on December 1, 1984. I came home with the intention of having a long conversation with Mom about family history and family life. This was a wonderful time to do it. Mom was 80. I was 40. We were 40 years apart, yet we were 40 years together. Although neither of us knew that we would only have six months more together, we did know that it was time to reminisce, about our past, for our future. This was the time to enjoy family conversations and family history, and yet preserve it for others. We were sitting at the kitchen table in the family house in Breda, listening (as we had often done in the past) to "Mantovani in Manhattan." We had often sat at the kitchen table over the years, just visiting or even playing Scrabble, one of Mom's favorite games. On those occasions, we used Scrabble to connect our lives in person while we connected our words on the board.

On this occasion we were talking family history. Talking about family and family life was always enjoyable, and I appreciated her comments. Mom's gentle demeanor had always belied her wisdom. Her natural shyness was quickly evident as I suggested that I wanted to tape record our conversation. She was hesitant. I put the tape recorder on the seat of the chair, out of her view. I put the small microphone on the table and then, before turning it on, started to talk about other things. We went through a number of old family photographs, and she identified as many as she could, sometimes by drawing sketches of the photograph and writing in identities and other times by making notations on the back. After we completed the photographs, we talked about our family, each of her children.

Mom was obviously proud of each of her children and she continued, as always, to be proud of each of them without showing any favoritism. It was a special quality of hers, and I admired it. Mom appreciated our differences without being judgmental of our decisions and actions. As we talked, she spoke of the pain of the family deaths that she had already endured, but she always recognized these pains and sorrows as part of God's plan, even when we, perhaps, had difficulty comprehending or accepting that plan.

Alice, by birthright, paved (if not bulldozed) the trail for the rest of us, sometimes carrying the special obligations of the oldest, sometimes enjoying the special privileges. Mom gave a special chuckle as she talked of Evie being "Miss Fancy Pants" in one of the earlier photos of the kids. Don was quickly recalled from his early marriage days on the farm, still trying to get by without indoor plumbing. Ed, she noted with some pride, had a striking resemblance to her dad, Grandpa Goecke. "A die-hard battery with a heart of gold," was how Mom lovingly spoke of Myrna. She spoke of Sal's days on her farm with a large family in much the same way that Mom probably saw herself on the farm with her large family. Larry's name evoked comments about how sports had permeated his life from the very start and how sports had in many ways fulfilled his life. As for me, Mom recalled how proud Grandpa [Frank] Wilberding would have been if he had known that I became a lawyer. She identified Shirley with her musical talent, appreciating their mutual love of playing the piano and the organ. Mom enjoyed the opportunity to enjoy each of her children and the opportunity to share that enjoyment with me.

Mom was relaxed now. She was talking about people she loved and events she fondly recalled.   It was time to turn on the tape recorder. During our preliminary discussions she became almost oblivious to the recorder and we simply had a nice conversation. What follows in this book is a transcript of our ensuing conversation to which I have added some notes and some photographs. This annotated conversation with Mom is naturally rich in family history. But, more than that, <u>it is Mom</u> – speaking to us as only she could speak. You will be touched by her love of family, her love of God, and her love of connecting her life to the lives of those who came before her or after her.

Evelyn Goecke Wilberding & Merle Wilberding
April 13, 1985

CONVERSATION WITH EVELYN
WILBERDING
December 1, 1984

[The recorder was turned on with Mom talking about how she met Dad and where Dad was living at the time of their first meeting . From that point, she started discussing the Wilberding farms, which we commonly called the "East Place" and the "West Place."[1]]

Merle: When you met Dad he was on the West Place?

Evelyn: Uh-uh.

Merle: East Place?

Evelyn: They were all living on that East Place.

Merle: They were on the East Place?

Evelyn: Yeah, when I first met him, yeah.

---

[1]The East Place and the West Place were the two family farms, both of which were about 160 acres and located one mile west and three miles south of the Town of Breda, at the intersection of what is now known as Falcon Avenue and 150th Street, in Wheatland Township, Carroll County, Iowa. Anton Wilberding purchased the East Place (known in the official records as the Northwest Quarter of Section 36) from Joseph and Elizabeth Shefers on March 1, 1894, for $ 6,920 (Book 5, Page 431, Deed Records of Carroll County, Iowa). He financed the farm with a $ 3,000 mortgage from Connecticut General Life Insurance Company (Book 6, Page 462, Mortgage Records of Carroll County, Iowa). Apparently there was a title defect on 40 acres of the East Place and Anton Wilberding later received a Quit Claim Deed to those 40 acres (the Southwest Quarter of the Northwest Quarter of Section 36) from James & Martha Callahan (Book 6, Page 597, Deed Records of Carroll County, Iowa). The West Place was purchased on March 21, 1922, for $ 28,320 by Frank Wilberding from the estate of his in-laws, Bernard and Katherine Gebhardt Wessling (Book 21, Page 532, Deed Records of Carroll County, Iowa).

Merle: And Wesslings were on the West Place?

Evelyn: No, let's see. I don't think the Wesslings were on the West Place.
          I think that house was all gone.

Merle: Really, it was just the ground?

Evelyn: Uh-huh, but then pretty soon after that they started building that
          first new house and then . . . .

Merle: The one that's on the West Place?

Evelyn: The one that's on the West Place, yeah.   And then we had to live.
          There was just that little shack on that East Place. I always thought
          I was going to get that.[2]

Merle: Is that the old wash house that we tore down, that one?[3]

Evelyn: Yeah, but it was standing on the same foundation where the white
          house was on the East Place.

---

[2]This exchange between Mom and me on the East Place and the West Place is a good
example of the natural propensity of us Iowans to define and identify things on the basis of
compass directions. In his book, *The Lost Continent* (1989), Bill Bryson humorously recalls
Iowans and their fixation with compass directions: "[H]e was like most Midwesterners.
Directions are very important to them. . . . If there are two Midwesterners present and they both
witnessed the incident, they will spend the rest of the afternoon arguing points of the compass
and will never get back to the original story . . . . European cities, with their wandering streets
and undisciplined alleys, drive Midwesterners insane."(*The Lost Continent* at 15).

[3]The old wash house was on the East Place. I recall helping demolish it in the early
1950s. At that time, it had not been lived in or even used for many years, other than for storage
of some junk appliances and equipment. For years, the house had been infested with honey bees
which had left huge deposits of honey between the walls. The walls themselves were
constructed of plaster over thin wood laths. This demolition job was orchestrated – well, it was
more a commandeering than an orchestration – by our hired man, John Hoogestraat, who
directed the demolition by Don, Ed, Larry, and me. (While I was only eight or nine years old,
I still had to help tear down the laths, remove the old honey combs, and haul away the trash.)

2

Merle: Is that right?

Evelyn: Yeah.

Merle: And they moved it?

Evelyn: And they moved it down there and they took part of it off, but then
we always used it kind of for a junk place.

Merle: Yeah.

Evelyn: Yeah, and then we had to live in it really until that new house was
done. I was married then.

Merle: Is that right?

The "new house" on the East Place
(Note the "outhouse" on the left)

Evelyn: Plus that, I had to feed all those carpenters. They stayed for dinner every time and supper.

Merle: Is that right, that was in 1929, 1930?

Evelyn: Yes, uh-huh.

Merle: How long did you live in that wash house?

Evelyn: Yeah, let's see. That would have been, I know we was living there already when our first baby was born and stuff like that. I was pregnant and trying to cook for those darn carpenters and feed them in that little shack and . . . . Yeah, I was in the house when they moved it.

Merle: Were you?

Evelyn: Uh-huh.

Merle: Is that right?

Evelyn: They just had something, some contraption and just pulled it over there and you got . . . .

Merle: Did they have horses?

Evelyn: I don't remember. I think they had tractors.

Merle: Tractors. That would have been about 50 to 100 feet.

Evelyn: I think they had tractors, and I don't think they moved all of that house.

Merle: Okay, when you got married, where did Grandpa and Grandma live then?

Evelyn: Well, they was living on the new place.

Merle: On the West Place.

Evelyn: Yeah, yeah, yeah.

Merle: Okay, so you got what we used to call the wash house?

Evelyn: Yeah, uh-huh. I thought when they was building that West Place at that time that we would get that, but they moved into it and they liked it, I guess. And they deserved it.

Merle: Who was living with them then? Was any of Grandpa's brothers living with . . . .

Evelyn: Well, he had that, well, first he had a pretty good hired man and it was, John was his first name. And he was, oh, he was between 30 and 40 but he was a single fellow and he was a pretty good hired man. He came from Omaha or something where Grandpa seen him in an ad or something and got him out there. And so he lived out there until we got married. We was married in January and then I don't know if it was February or March when Grandma and them was going to butcher, and Tony had a gun and he was supposed to go over there and shoot the hog or whatever it was. And there was so much snow he was kitty-cutting across from our house to the corner like where the mailbox was.

Merle: Yeah.

Evelyn: And as he got to that fence, you know, he was going to crawl over it and he had boots on. He stuck the gun in his boot and when he pulled up the thing went off. And he shot himself in the foot. And yet he hobbled the whole way over . . . .

Merle: This was . . . .

Evelyn: Tony, the first year we was married.

Merle: Dad?

Evelyn: Yeah, that's what he done. And then he got over there and then
they decided they better get him to the doctor or something.
Anyway, they called me and they put him in the hospital for that,
and he was in the hospital quite a while and they got the bullet out.
But in the meantime, we had to have a hired man at our place
because we had cattle and stuff like that, so Grandma and Grandpa
sent this, John Decker was his name, now I remember, John Decker
over to my place and he stayed there then and did the chores there,
and that's when they hired old George Wessling.[4] He had all those
kids and stuff, you know.

Merle: The trouble-making kids?

Evelyn: Yeah, all those trouble-making kids, yeah.[5] But I don't . . . .

---

[4]George Wessling was a brother to Grandma Mary (Wessling) Wilberding. This George Wessling was born on February 9, 1869, and married Mary Berning on July 27, 1904. It is important to keep this George Wessling separate and distinct from his first cousin, Reverend George F. Wessling, the son of Joseph and Christina (Hoelter) Wessling. Reverend George Wessling was born on December 23, 1883, and studied in Mount Calvary, Wisconsin, at St. Lawrence College. He was ordained on June 5, 1909, at Saint Meinrad, Indiana (where Father Joe Wilberding would later be ordained in 1943). He died on February 19, 1956, and is buried at St. Mary's Cemetery, Auburn, Iowa.

[5]Those "trouble-making kids" that Mom is talking about include George Wessling's son, Albert C. Wessling and Albert C. Wessling's son, Gary Lee Wessling. Their primary notoriety stems from a series of articles in *The Des Moines Register* which included, a report that in 1958 Albert C. Wessling was sentenced to 15 years in the Iowa State Penitentiary on burglary charges. He was described by police at that time as a member of a safe-cracking gang that was operating in Iowa. Gary Lee Wessling was implicated in the killing of a Des Moines cab driver in 1958, and, in 1963, he was indicted for first-degree murder in the killing a Grinnell policeman. On January 3, 1964, he pleaded guilty, for which he received a sentence of life imprisonment (*The Des Moines Register*, January 11, 1964). At his sentencing hearing, Gary Lee Wessling expressed his hope to be able to rehabilitate himself.

6

The Bernard Wessling Family

[Mom's sketch to identify the Bernard Wessling Family Picture]

7

Merle: George was not Grandma's brother?

Evelyn: Yeah, George was Grandma's brother.

Merle: George was Grandma's brother.

Evelyn: Yeah, he was the brother to Gert's father, too. He was in that family.

Merle: Okay, all right, so Grandma's brother, a different brother was Gert's dad. Is that right?

Evelyn: Yeah, his name was Joe.[6]

Merle: Gert's dad was Joe.

Evelyn: Yeah, Gert's dad was Joe and this one was George.

Merle: Okay, and Joe was a brother to Grandma and George was a brother to Grandma.

Evelyn: Yeah.

Merle: Okay.

Evelyn: But George had those little kids, I remember they did come up once. I don't know whether they was living together all the time. But they always said she was kind of on the wild side. She wasn't settling down too good. They lived in Sioux City.

---

[6]Joe Wessling was another brother to Grandma Mary (Wessling) Wilberding. He was also the father of Gertrude, Catholeen, Agnes, Mick, Ben, Jim, Blondina, and Sister Louise, F.S.P.A. Joe Wessling was born on February 6, 1867; he married Teresa Tebben on October 22, 1895; and he died in 1929.

Merle: George did?

Evelyn: Yeah, at the time, but he came out there. I guess they needed money
          and he had a good place to stay then. So, anyway, then that whole
          year, George helped Grandma and Grandpa, and we had Decker
          over at our place. And then by June they started, or May, whatever
          it was early in the spring when they could start a carpenter, and then
          they moved that house to the side and built that other house.

Merle: Did Grandpa's brother live on the West Place then with Grandpa?

Evelyn: Grandpa's brothers?

Merle: Did Henry or August . . . .

Evelyn: Uh-uh. They never stayed with Grandpa at all, no.

Merle: What did they do, do you remember that, at the time?

Evelyn: Well, August, I remember him and that was before we got married,
          that he had been in the seminary. August wanted to be a priest.[7]

---

[7]August Wilberding was the son of Anton Wilberding and Mary Grever and was also
a brother to Grandpa Frank Wilberding. August Wilberding was born on December 22, 1883.
In his early years he attended a seminary in Milwaukee, Wisconsin. Later, he decided not to
pursue that vocation and returned home. After I had this taped conversation with Mom, I had
a second taped conversation that same day with Gertrude Wessling. I have had this
conversation transcribed also, and I will refer to it as the "*Wessling Tape*." In this conversation,
Gertrude provided some interesting insights to August Wilberding, or "Gus" as she called him.
Gertrude recalled that Gus had gone to St. Francis Seminary along with Father George
Wessling. Even more interesting, Gertrude recalled that her own mother said "to her dying day
that Gus had a vocation, but she said he needed encouragement and he didn't get it at all, and
when he came home, he just felt real bad; he was just lost" *(Wessling Tape* at 18). Gertrude
remembered playing cards with Gus as she was growing up. "He was a very quiet person and
didn't have much to say" (*Wessling Tape* at 18). That demeanor may have cost Gus his
vocation because, as Gertrude said, "you know how strict they were in those seminaries and if
he didn't have everything just top notch and could rattle it off, then you were down" (*Wessling
Tape* at 18.) (See Footnote # 21 for further discussion of the death of August Wilberding.)

Merle: August was?

Evelyn: Yeah, August wanted to be a priest. But he gave it up and he came
          home and I remember him coming to church with Grandpa and
          Grandma and Tony really before I married him even. He was living
          there and so was Grandpa's mother because I was working at Mrs.
          Wilson's. Anyway, when Grandpa's mother[8] died, she asked me that
          morning, she knew I had been going with Tony, I guess, if I wanted
          to . . . .

Merle: That would have been in 1926, yeah.

Evelyn: Yeah, and then if I wanted to go to church for that funeral. I said,
          yeah, I'd go, of course, I never seen her even in the coffin. I just
          went to church that morning. It wasn't that serious that I . . . .

Merle: Yeah. Okay, so August studied to be a priest for a while. Where
          would he have studied, in Dubuque?

Evelyn: Gee now, let's see. There was some stuff down in the basement
          there for a while. I don't know if I ever pitched it all or not. But
          there was letters and stuff.

Merle: How about Henry? Henry was the one with the birth defect.[9]

Evelyn: Yeah.

---

[8]Grandpa Frank Wilberding's mother was Maria Grever who was born on May 4,
1841, and married Anton Wilberding on July 6, 1864, at St. Peter & Paul Church in Holdorf,
Germany. She died on February 23, 1926, and is buried at the cemetery in Breda, Iowa.

[9]Henry Wilberding was the son of Anton Wilberding and Mary Grever, and was also
a brother to Grandpa Frank Wilberding. Henry Wilberding was born on July 7, 1872, in
Holdorf, Germany. He never married. During his adult life, he worked as a hired man on the
farm owned by William Hires. Henry Wilberding had a very unusual birth defect consisting of
a splotchy grape-like birthmark over the entire left side of his face. Henry died on March 28,
1952.

10

Merle: Is he the one who lived with Bill Hires at some point?

Evelyn: Yeah, he did.

Merle: Did he do it way back then or did he help Grandpa farm?

Evelyn: No, he, I never seen much of him at all.

Merle: Really.

Evelyn: Uh-uh.

Merle: I remember seeing him when I was real young that he came out to visit a couple of times.

Evelyn: Yeah, yeah, that he did. But, I mean, he never, he never farmed out there or anything, but the Hires kept him there and I think he done enough chores around that they kind of liked to have him there, too. I don't know, and then later on he moved to Breda with some other old man, and then he was getting worse and they had him in the nursing home there until he . . . .

Merle: You never knew Anton. Anton died way before you ever met Dad.

Evelyn: No, uh-uh.

Merle: When you met Dad, did he have a car?

Evelyn: Yeah, I think he did.

Merle: He drove out when you'd go out to the farm. You would go out in a car, as opposed to a horse and buggy.

Anthony Wilberding

Evelyn: Oh, yeah. I never went in a horse and buggy with him. No, he had an old car and then once he got a new car, I know, before we was married. It was a new Ford and he was pretty happy with that one.[10]

Merle: Did you have tractors or horses to farm? Do you remember?

Evelyn: Well, I think, I believe they had a tractor then already.

Merle: Plus horses.

Evelyn: Plus horses, too. But I remember him when he used to take me to dances and stuff before we was married he would talk about he had

---

[10]I was amused at this comment of Mom's. It was no secret in our family that Dad loved new cars, and he had lots of them over the years. I just did not realize that this love of cars predated Mom's marriage to Dad. To illustrate Dad's love of cars, I will just note that during the ten years of the 1950s, Dad owned a black 1951 Oldsmobile, a blue and white 1953 Buick, a grey and white 1954 Buick, a red and white 1956 Buick, and a red and white 1959 Ford.

12

a, oh, you know, what all that machinery stuff, especially that rake, drag rake, you maybe don't even remember it there. I mean they had a rake which they had horses hitched to. You'd have to walk behind it, up and down all day long.[11] And I remember . . . .

Merle: You mean to rake the hay together.

Evelyn: Yeah, to rake the hay or even, you know, when they'd plow and then to smooth it down they'd rake it to smooth it down.

Merle: Oh, like a harrow.

Evelyn: Yeah, a harrow, yeah, I guess that was it. That was what they really called it.

Merle: Yeah.

Evelyn: And I know he would come in and he'd say he had walked all day on that old harrow and stuff, you know. He could hardly dance no more. But, he'd make it there anyway.

Merle: Oh, yeah.

Evelyn: And then when Grandpa's, well we didn't get married right away then, must have been another year. When they moved to that new West Place, yeah, because I know old Alphonse Wolterman, he was a brother to Lawrence, they lived on that East Place then for a year before we got married.

Merle: Before you got married?

---

[11]Even as late as 1950, when I was only six, I could remember the two work horses, Dick and Prince, being used to pull the rope for putting the hay away in the barn. (See Footnote # 42 for further discussion of the hay rope.) Much of the horse-drawn machinery was still on the farm. Unfortunately, there were only a few photographs of these horses on the farm.

Evelyn: Yeah, before we got married, yeah, because I know Alphonse was married, and we used to chum around, he was a brother to Lawrence. And then Lawrence used to work out and help Alphonse out there on the farm and that's how he got acquainted with Florence.[12] Because some nights when the snow was so bad they couldn't come through the roads, Lawrence Wolterman and Tony would walk, they would go east where the Schroeders was, you know, and get on the railroad track and they would walk into Breda on the railroad track.[13]

Merle: Is that right?

Evelyn: Yeah, and they'd both come up, yeah, because Lawrence was there and then they'd both . . . . Well, Florence, I always thought was just a kid when I got married. But Lawrence kept after Florence after Dad and I got married, and in two years she was married. She was married at eighteen and I just thought she was a kid, but then Tony didn't want to walk the tracks alone neither so they would walk together and go home through midnight. Can you imagine?

Merle: Is that right? On the train tracks. Five miles out there.

Evelyn: Yeah, and then one night I know he said they saw lights on the railroad track and they both was a little scared. They didn't know

---

[12]Florence Goecke was Mom's younger sister. She was born on March 24, 1912, and married Lawrence Wolterman on January 20, 1931.

[13]This railroad track was the original Chicago & Northwestern Railway system – probably the very track on which the Wilberding family traveled when they first came to Breda, after arriving on the boat from Germany. In reviewing the centennial books for the Iowa communities of Breda, Carroll, and Dyersville, Iowa, it is clear that the extension of the Chicago & Northwestern – which extended from Dubuque and Dyersville in northeastern Iowa to Carroll and Breda in southwestern Iowa – was a major link for the development of Breda and a major explanation as to why so many of our relatives, upon arrival in the United States, lived for a time in Dyersville before heading farther west to Breda to establish their homesteads.

14

who was sitting out there, and they never did find out what those lights was.

Merle: Well, the first ten years you were married, you didn't have electricity did you? The first ten, twelve years?

Evelyn: No, not until after Larry was born.[14]

Merle: Yeah, what did you do for milk and butter?

Evelyn: Well, we churned our own butter and separated the milk. Got the cream.

Merle: How did you keep it cool?

Evelyn: You'd hang it down in the well, either that or down in the basement, but we had a well, you know, where you'd lift up . . . .

Merle: Yeah, you didn't have ice or anything.

Evelyn: No, we never had ice.

Merle: You just left it in the well.

Evelyn: Yeah, you put it in a pail or something like that and tie a rope to it, and then you could pull it up when you wanted, or even keep it in the basement for quite a while. It would stay cooler down there.

---

[14]Larry Wilberding was born in Carroll, Iowa, on July 23, 1941– in the heart of threshing season, a time Mom often described as the hottest time of the year. Mom often described that year as a watershed year in the sense that 1941 was the year in which electricity was extended to the farm and, as a result of that, she got her first refrigerator. It is almost impossible to imagine the effect on her daily life as she first enjoyed a refrigerator after almost 40 years of hanging milk, butter, and other perishables down the well.

The Wilberding Children in 1940

1940 Marilyn (about) 1½ years old
    Myrna   "   3
    Eddie   "   4
    Don   "   6

    Evie (MISS FANCY PANTS)   8½

    Alice   9½

16 & Brownie (the Dog)

[Mom's handwritten description on back of photo]

Merle: What about meat when you'd butcher?

Evelyn: Well, we packed that in big gallon jars and salted it down. You
           would salt it down for a while, you know, put salt on it and then
           they'd hang it up in the smoke house and you'd smoke it. Those
           hams was just like a smoked ham then what you'd buy.

Merle: I remember the little smoke house on the East Place.

Evelyn: Yeah, on the East Place, yeah.

Merle: So you just salt the meat down and hang it up and let it dry out.

Evelyn: Yeah, yeah.

Merle: What about chickens, do the same thing? Did you butcher chickens
           then?

Evelyn: No, just what you'd butcher as you'd use them. Catch them out of
           the yard.[15]

---

[15]Of all the images I remember from living on the farm from 1944 through 1959, the image of butchering chickens may be the most vivid. By this time, we had a big freezer in the basement and we were able to butcher a lot of chickens at one time and then freeze them for future use. This was a major family project, with each of us having his own little job to help. I must say that I don't think anybody had a "good" job. We used the stump by the chicken coop (the same stump Larry and I used for first base in our baseball games) as the chopping block. Their heads would be lopped off, and, after jumping around like the proverbial chickens "with their heads cut off," the chickens would be gathered up, plucked, scalded, and made ready for the deep freeze. While there are no photographs of this, there is a photograph of Shirley and me standing next to the stump, holding a turkey awaiting Thanksgiving as only a turkey can. Myrna has reminded me that Mom and Dad won this turkey about three weeks before Thanksgiving, playing Bingo at a "Feather Party" sponsored by St. Bernard's Church in Breda, and that during those three weeks, the whole family had adopted this turkey as our pet. But come Thanksgiving, he still got the axe.

Shirley & Merle Wilberding

Merle: Catch them, and butcher them, and eat them right away.

Evelyn: Yeah, uh-huh. No we never put anything like that, never had freezers or any way to preserve chickens even.

Merle: Did you butcher cattle or just pigs? Do you remember?

Evelyn: Both. Yeah, yeah, we did.

Merle: Would you butcher them yourself, you shoot them and . . . ?

Evelyn: Yeah, yeah, oh yeah, and then you'd hang them up in the corn crib overnight, you know, you'd have one of those pulleys, you know, and you'd hang them up way to the top and like they do in the meat plants.

18

Merle: But you didn't have anybody else come out and butcher them, you'd
          just . . . .

Evelyn: Oh, no. It was always done by neighbors or friends.

Merle: You took the hide off and gutted it and . . . .

Evelyn: Yeah, uh-huh.

Merle: Cut it up for whatever pieces you wanted.

Evelyn: And plus that you clean the guts and make the sausage. That was
          one of my jobs. Take it way out in the grove, you know, and clean
          them out and then you'd have to scrape them real good. You'd turn
          them inside out and then scrape to get all that . . . .

Merle: The intestines, you mean?

Evelyn: The intestines, yeah.

Merle: Okay, and then you used that for the casing for the sausage.

Evelyn: For the casing to make sausage, yeah.

Merle: Is that right?

Evelyn: Yeah, yeah, you'd do that out in the grove.

Merle: Just to clean it, you'd take a bucket of water out there?

Evelyn: Yeah, and then you'd have to clean it and then you'd always soak
          them in salt water overnight, but they would get just as smooth and
          clean and white, you know. But it was a mess. I don't know how
          I stood to do it. But, we always done it at our home too, really.
          When we was young we learned to clean casings. Clean rabbits and

all that stuff. I especially did a lot of that where they'd shoot rabbits, and we'd have to take the hide off and pull that off and clean them.

Merle: Did you have to help out in the fields? Drive the horses and . . . .

Evelyn: No, I never worked in the fields.

Merle: So Dad and Grandpa, and then you had two hired men sometimes.

Evelyn: Yeah, two hired men, yeah.

Merle: You probably never knew what they paid those hired men, do you?

Evelyn: I don't know what we paid those first ones, but I do know, when the banks was all busted and stuff, you could get hired men for nothing, just for board and room. Yeah. And there was that one guy that Tony, well, I don't know where they found him, I think they was advertising something in the Omaha paper, and Tony got one of them that they would work for board and room, you know. And, what was his name? Dave, was his first name, I know. Dave McAdams, that's who he was, and I got a picture of him holding Alice when I was in the hospital with Evie, because that was just at the time of the Lindbergh kidnapping and when they went to stop for gas, Tony said the guy said: "You haven't got the Lindbergh baby in there have you?"[16]

---

[16]This incident would have happened, of course, within a few days after Evie's birth on April 27, 1932. Alice, who was born on February 9, 1931, would have been about one year old. The Lindbergh baby (Charles A. Lindbergh, Jr., born on June 22, 1931) would also have been about one year old at this time. The kidnapping and subsequent murder of the Lindbergh baby was *the* news story of the time. The kidnapping had taken place on March 1, 1932, but the dead body was not found until May 12, 1932 – about two weeks after this incident being described by Mom. It was more than two years before Bruno Hauptmann was charged, convicted and executed for this crime. Charles Lindbergh was America's hero for flying a single engine plane, The Spirit of Saint Louis, in a solo transatlantic flight that began in New York City on May 20, 1927, and ended thirty-three and one-half hours later in Paris, France. (Mosley, *Lindbergh* (1976); Lindbergh, *The Spirit of Saint Louis* (1953).

Merle: Is that right?

Evelyn: Yeah, and they had Alice all dressed up and I suppose Grandma
          helped take care of her or something like that.

Merle: How about vegetables?

Evelyn: I don't know how much vegetables and stuff we had . . . . We kind
          of smooched on them and got a lot, I don't know what we would
          have done without them.

Merle: Grandpa and Grandma?

Evelyn: Grandma and Grandpa. Whenever we wanted to have potatoes, we
          never planted so many potatoes.

Merle: You didn't? I remember planting . . . .

Evelyn: We planted some, yeah, but Grandpa always planted a lot, like
          they'd get a wagon load and they'd have that one room in the
          basement towards that east side. It was kind of a small room.
          They'd have potatoes packed in that whole room, you know.

Merle: Well, I remember helping picking potatoes and putting them in a
          wagon and filling that room up.[17]

Evelyn: That could be. Yeah, that could be.

---

[17]In the basement on the West Place, there was a large "fruit room," part of which was
set aside for the storage of potatoes. Consisting of approximately 50 square feet, this potato
room had a trap door to the outside, through which the potatoes could be shoveled down a chute.
The potatoes themselves were grown on several acres set aside in one of the fields. With a
potato digger, the dirt would be turned over, exposing the potatoes. We would then walk behind
the wagon, pick up the potatoes and throw them into the wagon. The natural cool temperature
of the basement enabled us to store the potatoes year round without spoilage.

Merle: When we were little, I remember that.

Evelyn: Yeah, so we always had plenty of potatoes and meat, and we had
           chickens, and we'd have a lot of eggs and milk cows and had the
           milk and . . . .

Merle: Did you have a big garden for vegetables? Did you grow a lot of
           vegetables or . . . .

Evelyn: I never was much good at . . . but we had cherry trees, I remember.
           We'd pick cherries off of those high trees. They was bigger than
           what we had here.

Merle: Did Grandpa talk much about Germany in the early days?

Evelyn: Yes, just when I'd be real busy in the kitchen, he'd love to come in
           and sit by the stove, you know, in the winter time, and he'd tell them
           how it was all in Germany, you know. And I'd say, oh, yeah, yeah.
           Go in one ear and out the other 'cause I had always something to do
           with the kids and everything, and at that time I didn't give a darn
           what Germany looked like. I was trying to get my own kitchen
           cleaned. So I wish now many times that I had really sat down and
           had it soak in really what he did tell me.

Merle: What did he tell you about coming over?

Evelyn: Well, that I remember him telling about . . . they went to the bottom
           of the boat. They came across on a boat. And there was some old
           people there. He said they were like real old people and they didn't
           have anything, and some of them got sick and was puking over the
           sides. They got seasick, and they had to stay down there because
           they couldn't afford to buy a place on the upper part of the ship.

Merle: On the upper decks, they were on the bottom.

Evelyn: Yeah, they were on the bottom of the deck he said, where they had old cattle and stuff like that sometimes on there, too, close by, but not just in the same room, I guess. But I know he said there was some really poor people that were sick on there, too.

Merle: Yeah, yeah. Did he talk about anybody ever dying coming over? Anybody on the ship?

Evelyn: No, not that I ever heard.

Merle: What about when he landed in Baltimore? Did he ever say what happened there? The first week or two weeks or whatever it was?[18]

Evelyn: I don't know how they got out of Baltimore so fast or whether . . . I think, let's see what Marie Schumacher's dad . . . . [19]

Merle: Charlie?

Evelyn: Charlie, Charlie was kind of helping them to get out this far because they came out here then. They just headed right for here because Charlie had been here for quite a few years, and that's how they come way out here to Iowa.[20] And Grandpa wasn't married at the

---

[18]I have a recollection of Grandpa Frank Wilberding sitting at the kitchen table on the West Place and telling us that when they landed, they were penned up like pigs for two weeks before they were free to go. While Mom and I were talking about Grandpa Frank Wilberding's arrival in Baltimore, I have since found some evidence that this arrival may have been at Ellis Island in New York, but this research is still continuing as of date of publication of this book.

[19]The U.S. Passenger Records show that on May 29, 1890, Charlie Wilberding arrived in Baltimore, Maryland, on the *S.S. Dresden*, a steamship carrying 1,138 passengers, almost all of whom were in steerage. (National Archives Records: Index: QQ-1-7-372; Manifest Microfilm M255, Roll 47).

[20]Charlie Wilberding was born on April 9, 1865, in Holdorf, Germany. On November 15, 1909, he married Minnie Fink. Charlie Wilberding died on December 19, 1936. Their children include Father Joe Wilberding, Agnes Wilberding Flies, and Marie Wilberding Schumacher.

time, you know, so they lived there, I think, with Charlie, and I don't know how they got on that East Place but I know that Grandma was on the East Place with them. And so what that one that was studying for a priest, what I'd say he was?

Merle: August.

Evelyn: August, yeah. He was studying to be a priest and whether he was still had the priest vocation, I don't know, but anyway he came home and I remember seeing him come in church with the Wilberdings, but then he died, too, before we was married.

Merle: Which one, August?

Evelyn: August, yeah.

Merle: Yeah, he died in 1925 or 1926, somewhere in there.[21]

Evelyn: Yeah, he died just before we got married probably.

Merle: Did the Wilberdings from Minnesota, did they go to Minnesota right away or did they ever live around here? Was it Ben Wilberding?

Evelyn: Yeah, that was Ben Wilberding. I don't know. They lived in Minnesota as long as I knew them.[22]

---

[21]August Wilberding died on January 20, 1926, after a short illness of double pneumonia. It was noted that for the few years prior to his death, August Wilberding had worked for his brother (Grandpa) Frank Wilberding. In his obituary, August was praised "as a man of quiet disposition, earnest and industrious, and will be greatly missed by all." (*The Breda News*, January 28, 1926). (See Footnote # 7 about August's attempt to become a priest.)

[22]Ben Wilberding was the son of Anton Wilberding and Maria Grever. He was a brother of Grandpa Frank Wilberding. Ben Wilberding was born on December 24, 1874, in Holdorf, Germany. On February 19, 1901, he married Agnes Fehring in Mt. Carmel, Iowa. On March 19, 1950, he died and was buried in St. Anthony's Cemetery in Lismore, Minnesota. The Ben Wilberding family lived in Lismore, Minnesota, for many, many years.

24

Merle: Did they?

Evelyn: Yeah, just when they went up there, I don't know. There's one that
          married Rose, she was Rose Wilberding, married Joe Wittry. You
          remember back yonder, their little girl was flower girl along with
          Myrna at that time.

Merle: Who were the Wilberding priests besides Father Joe? There were
          other Wilberding priests.

Evelyn: Yes, I remember Father Joe.[23]

Merle: Who were they?

Evelyn: Well, that I don't know.

Merle: I saw one in one of those obituaries that there was a different Father
          Wilberding said a mass for, I forgot which one, maybe it was
          Charlie, in fact, but it was not Father Joe. It was a different one.

---

[23]Father Joe Wilberding was the son of Charlie Wilberding and Minnie Fink. Father
Joe was born on December 28, 1910. He graduated from Loras College and was ordained in
Saint Meinrad, Indiana, on June 5, 1943. His First Solemn Mass was on June 15, 1943, at St.
Peter & Paul Church in Carroll, Iowa, at which Father Flanagan, the founder of Boys Town in
Omaha, Nebraska, gave the homily. (Father Joe had worked at Boys Town during his student
days.) He served at a number of Catholic parishes in Missouri, died in Columbia, Missouri, on
September 23, 1980, and is buried at St. Mary's Cemetery, in Willey (Carroll County) Iowa.
(I remember Father Joe as someone who loved to play the piano, as he often did at family get-
togethers over the years.) Another Wilberding priest was Father Herman "H.A." Wilberding
from Haverhill, Iowa, who was the son of Herman Wilberding and Mary Heintz. Herman
Wilberding (H.A.'s father) and Grandpa Frank Wilberding were first cousins. Father H.A.
Wilberding was born on January 21, 1883, was ordained at Dubuque, Iowa, on July 2, 1907,
and died on September 29, 1938. Father H.A. Wilberding officiated at the funerals of, among
others, Great-Grandmother Maria Grever Wilberding in 1926, and Charles Wilberding in 1936.
He served as Pastor of the Immaculate Conception Parish in Haverhill, Iowa, from 1926 until
his death in 1938 (*Haverhill, Iowa, Centennial 1882-1982*, Pages 57-58).

Father Joe Wilberding

Evelyn: Father Joe. I always remember there was another Father Wilberding at Joe's when he was celebrating his First Mass. Yeah, when we was at the St. Angela Academy. That was just across from where Charlie and Minnie lived on the corner, you know. That house is torn down now.

Merle: Is it?

Evelyn: Oh, yeah, for quite a few years. But anyways, it was just across the street there and we thought that was so great, and they had all the dinners and all the relatives there. But I know there was another Father Wilberding in the relation, but where he went to or come I don't know but . . . .

Merle: Did Grandpa ever correspond or write the people in Germany that you know of?

Evelyn: I remember him, I think he sent him money once, probably it was some letter that he had got from them that they needed help like that one you was reading about that.[24] And that might have been one, because I know he was very good-hearted to give anybody something, but he was tight on himself. Of course, they . . . .

Merle: Yeah, he didn't spend anything on himself.

Evelyn: Not if they didn't have to, no.

Merle: Yeah, I was probably only 10 years old and I would remember him having his books out and figuring out interest . . . .

---

[24]Grandpa Wilberding had retained this letter he had received from one of our German relatives, the Louis Espelage Family. (The writer identifies his wife's sister as Katharina Strunk who was a cousin to Grandpa Frank Wilberding. The Strunk connection to the Wilberding family originated on July 21, 1858, when Regina Catherina Wilberding – a sister to Anton Wilberding – married Bernard Heinrich Strunk. This Wilberding-Strunk line has stayed in Germany and includes Josef Lünneman and Maria Blöemer with both of whom we have rekindled our family relationship.) In his letter of September 9, 1947, he was pleading for food because their plight after World War II was so dreadful. Some excerpts of the letter are as follows:

> When the war was over, we . . . didn't know anything about [the whereabouts of our children but] stayed with a farmer in his attic because our home was totally destroyed from the last bomb attacks. . . . Potatoes, we will get 100 lbs. from now until next harvest per person. And almost no meat at all . . . . It looks like we have a winter of potatoes and bread ahead of us . . . . We used to have our own pig, but where we live now, we don't have a barn and it would be stolen anyway. By the end of the war, we still had 15 chickens, but they were all stolen. We brought 20 chicks from Oldenburg, but one day somebody took all of them. Now we are paupers, living on food cards.

Mom had commented several times over the years that Grandpa Wilberding had sent food and other things, including money, to help their situation.

Evelyn: Yeah, I think I pitched the book now because Tony and him are both gone but I know we had the book here for quite a while. Heck, he had a little notebook, you know. How many people in this county that he didn't loan money to?

Merle: Is that right?

Evelyn: That is true.

Merle: No kidding.

Evelyn: Yeah, there was a lot of them from Carroll and stuff, you know. If they'd buy a house, they'd come to Grandpa, and sure borrowed.

Merle: Yeah, he seemed to know where the dollar went pretty good.

Evelyn: Yes, he did.

Merle: I can remember him sitting at the table, too, talking about . . . .

Evelyn: Uh-huh and Grandma always said Dad gets so mad and I like a little cream in my coffee. And he thinks we've got to sell all the cream when they separate. He didn't even want her to use cream.

Merle: Is that right?

Evelyn: Yeah, they should just use milk or the skim milk for themselves.

Merle: What would they sell? Would they sell their corn? How would they get the money when they were farming? Would they sell corn and the other crops?

Frank Wilberding disking the cornstalks

Frank Wilberding riding the bobsled

Frank & Mary Wilberding

Evelyn: Yeah, yeah. He'd sell corn.

Merle: And the hay? Did he raise cattle to sell or just to eat?

Evelyn: Well, they had milk cows and stuff like that, and, no, I don't think he had big droves of them. They had milk cows, which Grandma would always get up in the morning to go down to milk the cows. She loved to work outside better than inside, really.

Merle: Did she really? I've seen pictures of her like on the big corn wagon, picking corn.[25]

Evelyn: Yeah, oh yeah. I never picked corn after I was married, but before I was married when we was at home, we always had to pick corn. You'd get up at 5:00, 6:00 o'clock and you'd pick 'til noon. And

---

[25]While there are precious few photographs of Grandpa and Grandma Wilberding, there are several wonderful photographs of them harvesting corn, disking the cornstalks, and riding the bobsled.

30

then once when we had ours out, Uncle Henry Otto, he was my uncle, he didn't have his out and Clarence, and golly we were just little, he lived closer to Lidderdale and we lived near Mt. Carmel where that old dirt place was, and we'd take the wagon and team and heck I bet we wasn't more than twelve years old, and go over there. And we'd stay overnight for the week out there, and Aunt Annie would feed us at noon, and we was always so hungry. I know when we'd come in from picking corn and it was cold, but we'd go on, you could get quite a bit of corn picked. I don't know just how much. Then they'd weighed it up in some place, and we'd get credit for what we picked, and we made money that way.[26]

Merle: What did Dad's brother die of? Little Alfred? Do you remember how he died?

Evelyn: He died before I knew him but then Tony always said and Grandma did too, he had kind of a stomach flu is what she said.[27]

Merle: Really?

Evelyn: Uh-huh and I don't know if we've got the picture of little Alfred. I think we've got some small ones. There was a big picture which we had in the attic first that I didn't hang up here, but he had dark hair

---

[26]When Mom talks about picking corn, she is talking about the manual process of harvesting the ears of corn from their stalks. Obviously, it took a long time to grab the ear off of each corn stalk and throw it in the wagon. As big a chore as that seems now (and probably then too), the harvest was not yet complete. Each ear of corn was still wrapped within its husk and this had to be removed by hand, usually with the help of a metal hook which was attached to your hand with a leather strap. But, even the earlier mechanical corn pickers did not eliminate the manual work. Even in the early 50s, the farm kids were let out for one or two weeks of "corn picking vacation." During this "vacation," we walked along side a wagon and picked up off the ground all the ears of corn which the mechanical corn picker had missed or discarded.

[27]Dad's younger brother (and his only brother), Alfred Wilberding, was born on June 23, 1912, and died on August 31, 1913.

and brown eyes, and Grandma, she always thought he was the prettiest one, which he probably was because she has brown eyes, Grandma. But none of our kids ever inherited brown eyes. But that little Alfred had brown eyes. And he would, he could creep, well, I don't know. I've thought of that different times, too. Now, what did he really have? But she said he had the summer flu, is what they called it, and they didn't take kids to doctors and specialists like they do nowadays. If they had anything they'd try to cure it themselves or get some medicine from the doctor but they didn't look into anything else.

Alfred & Anthony Wilberding

Merle: Did Grandpa ever say why they left Germany?

Evelyn: Well, because Charlie wanted them to come across, and they was
           poor in Germany. He said they was poor.[28]

Merle: Did he ever say why Charlie came over all by himself?

Evelyn: Yeah, that I, he just wanted to venture out.

Merle: Charlie's obituary said that he served in the German army.[29]

Evelyn: That could be. Any maybe he thought that he'd come to America,
           where there would be peace and stuff like that and see what it
           looked like, and they brought the family over.

---

[28]During the period 1870 to 1890, a significant percentage of the population of
Holdorf, Germany, emigrated to the United States because of poor economic conditions. In my
notes of a conversation I had with Dad on January 3, 1969, he told me his understanding was
that, besides the poor economic conditions of Holdorf at the time, Anton's family was also
fearful of being conscripted into the German military forces.

[29]Charlie Wilberding died on December 19, 1936, and had the following obituary:

Charles Wilberding was a son of Mr. and Mrs. Anton Wilberding, early
residents of the Breda vicinity, and he was born April 9, 1865, in Oldenburg,
Germany. He grew up to manhood in his native country and served three
years in the German army. In the year 1890, Mr. Wilberding left his native
country and came to the United States. He first located in Dyersville and
later came to Breda.

He engaged in farming a short while after his arrival here and on
November 15, 1909, he was united in marriage to Miss Minnie Fink. The
ceremony was performed by Rev. G. F. Wessling at Sts. Peter and Paul's
Church in Carroll. Following their marriage, they resided on the farm for a
time and then moved to Carroll in 1913 which has since been their home.

Three children were born to Mr. and Mrs. Wilberding who are Rev. Joseph
Wilberding, and two daughters, Misses Agnes and Marie Wilberding who
reside at home.

Merle: Did you ever visit the Wilberdings in Dubuque? Are they, were they
           relatives?

Evelyn: Yeah, they was relatives, and the only time we did was once when
           we went to, I don't know, if we would have went up to where you
           was at in Minnesota, and we went around there to the Mississippi
           somehow, and then we stopped at Dubuque. And Tony was going
           to look up some of those Wilberdings. There was a big store that
           had Wilberdings.

Merle: So you stopped in the Wilberdings' store in Dubuque?[30]

Evelyn: Yes, there was a lady there.

Merle: Was she a Wilberding?

Evelyn: She was a Wilberding. Whether she was married to a Wilberding,
           why I don't even remember that, but anyway her name was
           Wilberding and she worked at that store and she took us out to
           lunch.

Merle: Did she?

Evelyn: Yeah, she did. She said we should come along, she'd take us out to
           lunch then.

Merle: Did you ever visit them back 30 years ago, 40 years ago, or 50 years
           ago when you were growing up? Did the Wilberdings from
           Dubuque ever come here or vice versa? How were they related?
           Did Grandpa ever say? I know when Grandpa first came here they

---

[30]The Wilberdings in Dubuque consisted of the families of Henry Wilberding and
Herman Wilberding, who were both sons of Carl Wilberding and Mary Ann Johanning. Since
Carl Wilberding and Anton Wilberding were brothers, Henry Wilberding and Herman
Wilberding would have been first cousins to Grandpa Frank Wilberding.

stopped in Dubuque or they lived in Dyersville for six months or a couple months or something before they came to . . . .

Evelyn: That could be.

Merle: That's what the obituaries talk about.[31]

Evelyn: That could be. We was up to Remson for some funerals in the early days of our marriage. Yeah, we went up there for some funerals, and I can't just think, but I do remember meeting the Wilberdings, of course, they were like cousins.

Merle: Did he ever talk about relatives in any part of the country in the United States besides the ones in Remson and Minnesota? Like in Ohio or the East or anywhere else?

Evelyn: No.

Merle: He never talked about that?

Evelyn: Not that I remember.

Merle: Did you eat many pheasants from hunting?

Evelyn: Oh, yeah.

Merle: Did Dad hunt a lot of pheasants?

Evelyn: Yeah, yeah, we used to eat pheasants. They was pretty nice if they wasn't shot up too badly. Then it was a mess.

---

[31]The German newspaper obituary for Great-Grandfather Anton Wilberding states that in August 1890 the family Wilberding emigrated to America and settled first in Dyersville, Iowa, but that the following year, they moved to Carroll County, Iowa. Great-Grandfather Anton Wilberding's only brother, Carl Wilberding (1827-1888) had lived in Dyersville, Iowa, and is buried there at St Francis Xavier Cemetery.

Anthony Wilberding

Merle: Did Dad ever do any tracking for muskrats or mink?

Evelyn: Oh, yeah, he did a lot of trapping for muskrats.

Merle: Did he? Fox or . . . ?

Evelyn: No, it was muskrats and then we . . . .

Merle: Along the creek?

Evelyn: Along the creek, that was when we was on the West Place already, you know, and then those Schwabe boys there, Paul and Roger. Well, Roger maybe was kind of small, but Paul . . . .

Merle   Larry and I had the same experience when we were young and used
           to trap along the Old Lady Creek in early November.[32]

Evelyn: Oh, did you?

Merle: Yeah, yeah. We used to trap them and take them down to the
           basement and skin them and put the pelts on those peach crate
           boards and then stretch them out and get them all hanging up there
           'til they were hard and then sell them. I can't remember where we
           sold them. I don't know if we mailed them off to some place or
           brought . . . .

Evelyn: I believe there was somebody in Carroll that would take in those.

Merle: Yeah, I kind of think so, too.

Evelyn: It seems like I read where that muskrat stuff is pretty expensive right
           now if you . . . .

Merle: I don't know, I saw last week in Ohio that muskrat pelts only bring
           about $3.50 a piece.

Evelyn: Oh, is that so?

Merle: Which isn't very much. Because we used to get $1.15 and $1.20 for
           them thirty years ago. So, they are expensive when you buy them,

----

[32]These trapping days might romantically be described as "Muskrats in the Mist," and,
in many ways, they were. We would get up at 5:00 a.m. and make the three-mile walk in the
crisp November air to check our traps, which we had previously set along the Old Lady Creek
that meandered through the pastures on both farms. Often, the moon would be shining and the
mist would be rising from the creek as we walked through the pasture, anticipating whether our
traps would yield any muskrats for us to grab and put in our gunny sack, for later skinning the
pelt and mounting it on a peach crate board which we had slightly trimmed to stretch the pelt,
prior to sale.

but like everything else, it's not much when you start out. How about that '36 snow storm? Wasn't that when Ed was born?[33]

Evelyn: Yes, that was terrible. And it was in February. Pete was, I don't know if it was Pete's birthday or what. Anyway, they lived out on 71,[34] on that, where the Oswalds lived, and, yeah, I think it was Pete's birthday.[35] Anyway, he was having a party over there, and Dad thought well he could make it through the snow. We had a real good hired man that time and so Tony took the car and then when he got halfway, he went around, I don't know where they went around the highway, I think, like they lived on 71, and I believe he took that to come in from the South and then he got stuck in a snow bank there around where those Grimsmans once lived, south of our place yet. And he had to walk home and, Lord, it was about 3:00

---

[33]The 1936 snow storm is still regarded as one of the worst snow storms of the century in Iowa. The snow storm was accompanied by winds of 40 miles an hour and temperatures reaching 25 degrees below zero. Heavy drifting of snow crippled all auto and train traffic and brought communities to a standstill. Schools were closed for a week, while emergency crews tried to bring in emergency fuel and food. Livestock and machinery alike froze in the sub-zero temperatures. The following account captures the effects of the storm and provides a good backdrop for Mom's bobsled ride over the snow-drifted fields and frozen creeks to town, so that she would be at the hospital to deliver Ed, who was born on February 26, 1936:

> The winter of 1936 was the worst in Iowa's history. Beginning on January 18 and running over to February 22, a blanket of cold settled over Iowa which held the temperatures below zero – at their warmest – for days at a time. Headlines would read: "MERCURY CLIMBING - TOWARD ZERO. - 8 AT NOON! MERCURY GOES ABOVE ZERO FOR THE FIRST TIME IN A WEEK!" Schools closed. Trains stalled. Traffic was at a standstill, even in cities. Towns ran out of coal and even food. Livestock froze and throughout it all, one great blizzard raged after another for more than a full month. (Hamilton, *Pure Nostalgia* 182 (1979)).

[34]When Mom is speaking of "71," she is referring to U.S. Route 71 which is the main north/south road to Carroll, Iowa.

[35]Pete Oswald was born on February 4, 1894. On November 23, 1920, he married Teresa Goecke, a sister of Mom's.

38

or 4:00 when he got home, and he was just a puffing and it's a
wonder he made it through there. And he said he couldn't see the
fence or anything, it was getting, we always had snow and then it
snowed so hard it was blowing and everything. You know how
those storms are? He managed to get home but he was just all in
when he come into the kitchen and was all snow, but he had to leave
his car there. And he had to walk that last way, and he didn't know
if he was on the road or in the fields or what. It's a wonder he made
it home.

Merle: Right. This was before Ed was born?

Evelyn: It was before Eddie was born, yeah. And then, well, when it got,
Eddie was due at the last of February, and when it got close to that
day, well you couldn't get out with anything but a sled, and one day
it got kind of nice, you know, so he got all the neighbors together
with horses and sleds. So we had two set of horses and two sleds
if one of them would act up, and all the neighbor men went along
and took me into town. And we decided that I was going to stay in
Breda because it was snowing pretty bad, and I stayed with
Clarence that night. Probably Dad stayed there, too. No, I don't
know where he stayed. I can't remember if he stayed there but the
Goecke girls still talk about it, that evening. I think he and the men
went back home again.

Merle: Uh huh.

Evelyn: They got home. But they had to go through because, Lawrence
Nieland reminds me of it, they came through, they cut kitty-corner
across the fields, you know, in the snow through the Nielands, yeah,
where they lived, yeah, because you couldn't get through the roads
and everything was so. . . .

Merle: So they just went across...

Evelyn: They crossed wherever they could get through over the fences and everything.

Merle: No kidding.

Evelyn: And they got me to Breda and then went up to Clarence's, but we couldn't get to Carroll. Those roads was all closed that night, but I felt good to get to town where there would be other people. And, so the next morning Tony Wand was going to go to Carroll; he was the mailman then. And he was going to go into Carroll when they had the roads partly cleaned. And he took us up to Carroll, and I stayed then for a few days with my uncle Frank, Frank Goecke and his wife Kate. They didn't have any children. And we stopped up there. I didn't want to go to the hospital. They probably wouldn't have taken you in there anyway. Then I was there a couple nights before Eddie was born.

Merle: Is that right?

Evelyn: Yeah, and then we had to go to the hospital.

Merle: When you were going in the bobsled, how did you keep warm?

Evelyn: Oh, we had plenty of coats and all those blankets and stuff. You didn't just get so cold. They always had those hay things that you would sit on. Those hay bales, yeah.

Merle: Did they have those hot rocks or bricks?

Evelyn: I don't know whether we used them then, but when I was a kid and went to school Mom would always have a hot rock in the oven, and we'd drive a pony, a blind one usually is all we got.

Merle: A blind horse?

Evelyn: To go to Mt. Carmel, a blind horse, yeah, old Flory. It was she couldn't see but she was very gentle, and we'd drive her and just a little surrey. No, it wasn't a surrey it was just a one seated thing, I think. And us kids would pile in there, and you would put that brick on the floor.[36]

Merle: On the floor?

Evelyn: Yeah, and you'd still be cold by the time you'd get there.

Merle: So, you went cross country in the bobsled.

Evelyn: Yeah, yeah, I did. And then, well before I went to the hospital, then we thought we had to have a hired girl in there, because there was Alice and Evie and Don. Little kids at home and they was small yet. So we, Tony asked around where there would be a good hired girl, and that's when we got that girl from, oh, she was from Arcadia. I'm so forgetful on names now, but then anyways, I never met her before, but they said she was such a good girl. What was her name?

Merle: I don't know. I don't remember. I know the girls would know, but I don't remember.

Evelyn: Oh, it will come to me eventually, but I'm just . . . .

Merle: When you had Ed in that snowstorm, and Dad got the neighbors together, did you have a telephone then?

Evelyn: I imagine we did.

---

[36]Mom had often talked about how they would heat a big brick or stone on the cook stove or in its oven, and then place it on a special platform in the wagon, the bobsled, the surrey, or whatever horse drawn vehicle they would be taking at the time. She also talked about and still had this very heavy "horse blanket" which was then thrown over the hot brick and huddled up to the people in the vehicle to contain the heat and direct it to their bodies.

Merle: You did not have electricity but you had a telephone?

Evelyn: We didn't have electricity. We had telephones, yeah. We used to
          have telephone meetings at our place. Tony was one of the
          presidents or the bookkeeper or something.[37]

Merle: Way back then?

Evelyn: Yeah, and they'd all come over there and you'd have to pay your
          dues for a year, which wasn't very much at that time.

Merle: And you were on a party line with, I guess, Grandpa then, too.
          And Gert and Schwabes.[38]

Evelyn: Yeah, and the Grimsmans were on that.

Merle: Who lived in that Freddie Freeze place then? Do you remember?

---

[37]Telephone service came to Breda in 1905. Then, there were multiple telephone companies in Breda, but these companies got together and formed the Breda Switchboard Company, which acquired and operated the Central Switchboard (*Older Days Renewed, Breda Centennial 1877-1977*, Page 275-277 (1977). (I will refer to this book as the *"Breda Centennial Book."*) For years, these multiple telephone companies continued to operate. One company served the farmers; the other company served only the Town of Breda. Eventually, in 1966, The Breda Farmers Coop, the Breda Switchboard Co., and The Breda Telephone Co. all merged into The Breda Telephone Corporation and then, for the first time, adopted the "dial" telephone system.

[38]During these years, anyone living on a farm had to share a "party line," which meant that multiple farms, usually six to eight, shared the same telephone line. There were about six separate farms on our party line, known as Line # 33. Our own phone number was "23 on 33" which meant that if you wanted to call our farm, you physically cranked the ringer on your wooden wall telephone. This would connect you to the Central Operator, and you then asked for "23 on 33." The Operator would connect her main switch to Line # 33 and then ring two long rings, followed by three short rings. Of course, all six parties on Line #33 heard those same rings and knew that the Wilberding family had a telephone call. And it gave each of the other parties the opportunity to "rubberneck" and listen to our call. Consequently, during those days, all of our telephone calls tended to be very short and impersonal.

Evelyn: Well, Freddie . . . .

Merle: Had Freddie lived there a long time?

Evelyn: Freddie lived there quite a . . . oh, you mean, who's living there
          now?

Merle: No, no. Who lived there in the 30s? Did Freezes live there then?

Evelyn: No, I think the Sieves was living there.

Merle: Sieves?

Evelyn: Yeah, old Joe Sieves. And they're both dead.

Merle: Yeah, yeah.

Evelyn: And then the Freddie Freezes come.

Merle: Did you have to help on the threshing and the hay?

Evelyn: I never did.

Merle: Just except for making food.

Evelyn: Making food and bringing out lemonade and sandwiches and stuff
          like that in the afternoon and the morning before dinner and then
          have dinner for them all in the house and . . . .

Merle: How many would there be?

Evelyn: Oh, fourteen.

Merle: Really, who owned the threshing machine?[39]

Evelyn: I think old Wesslings did.

Merle: Did you?

Evelyn: Uh huh.

Merle: Is that Thresher Joe?[40]

Evelyn: Yeah, that was Thresher Joe. He always had that.

Merle: Is Thresher Joe, is that Gert's dad?

Evelyn: Yeah.

Merle: Okay, I saw an obit where he was referred to as Thresher Joe?

Evelyn: Yeah, he was.

Merle: He probably owned it then.

---

[39]I have used the term "threshing" here to be consistent with the spelling of the name for a relative we called "Thresher Joe." Having said that, I need to point out that Mom generally used the term "thrashing." I looked to the dictionaries for help. Both terms are used, interchangeably, in the older dictionaries (*Webster's New International Dictionary* (1909 Edition)), but "thresh" is the preferred use in the more recent dictionaries (*American Heritage Dictionary* (1992 Edition)). In the nostalgia books, both terms are used. (Weitzman, *Thrashin' Time* (1991); Kammerude/Garthwaite, *Threshing Days* (1990)). In any event, the term is intended to describe the beating the grains (e.g., oats, wheat, or rye) from the stalks or straw.

[40]"Thresher Joe" Wessling was a son of Bernard and Katherine Gebhardt Wessling and was a brother to Grandma Mary Wessling Wilberding. Formally christened Joseph Francis Wessling (1867-1929), Thresher Joe was the father of Gertrude, Catholeen, Agnes, Mick, Ben, Jim, Blondina, and Sister Louise, F.S.P.A. He was known to his neighbors as "Thresher Joe" because for many years he owned and operated a threshing machine and threshed grain for many of the neighbors, during a time when threshing was both a cooperative work effort and a community social gathering.

44

Thresher Joe Wessling & Threshing Crew

Evelyn: I can't think who else would have taken over. And maybe somebody
did, though, too. I think Thresher Joe would have been . . . I think
he died early in our first year. But when Larry was born I know we
had thrashers that day and I had dinner ready for all of them and had
to call the . . . .

Merle: Thresher Joe died the year you got married?

Evelyn: Yeah, that's what I thought. I remember seeing him in bed there and
it was one of the first years I know. Maybe it was Muggenbergs
that had that thrashing outfit then, too. I don't know.

Anthony Wilberding using the hay buck

Merle: Okay, the threshing was for the oats and the straw. What about the hay? How would they do with the hay? Did they have . . . ?

Evelyn: Yeah, yeah, well, that they would, I think they would just go out there and mow it down with . . . .

Merle: And then rake it? What was a hay buck? Didn't they use a hay buck?[41]

---

[41]There were several different variations of this farm implement, but all were designed to help a farmer to make a haystack when the haystack itself was the problem – how do you put hay on top of a hay stack when the hay stack itself is getting higher and higher? (Klinkenborg, *Making Hay* 94 (1986).) Sometimes these were call buckrakes; sometimes these were called beaverslides – a name derived from its original patent title, the Beaverhead County Slide Stacker. We simply called them "haybucks" and the photograph in this book shows what our hay buck looked like it. The hay stacks made by the hay bucks were all "loose" hay.   Once we started putting the hay in bales, the haybucks no longer served any purpose and they were no longer used.

46

Evelyn: Yeah, if they had one of them they could roll it into a roll that way.

Merle: And then get it on the wagon of some kind?

Evelyn: Yeah, yeah, uh huh.

Merle: I remember when we used to bring the hay up with Dick and Prince. We had those work horses.

Evelyn: Oh, yeah. Oh, yeah. And Grandma used to like to drive the horses on that thing. "Go ahead," and then "Stop," you know, when the hay was ready to be tripped into the barn.

Merle: I remember them pulling the rope off in the barn to pull the hay up,[42] and I can vaguely remember Grandpa driving the two horses with, like, a little mower.

Evelyn: Oh, yeah. Uh huh.

---

[42]The hay was hauled up into the barn by a rope pulling a series of four metal L-shaped forks stuck into an interlocked stack of sixteen bales. Sometimes this device was called a grapple horse fork (Yale, *While The Sun Shines* 40 (1991)). "The fork was attached to one end of the hay rope [and the rope] ran up to the peak of the open haymow door, the full length of the barn, down the back of the barn, and up to the front of the barn." (Hamilton, *In No Time at All* 93-94 (1974). From there, the rope was connected to Dick and Prince, our two work horses, and these work horses pulled the rope, as first I remember it. Once the fork full of hay was pulled up and into the barn, it would ride down the track until the person in the haymow yelled "Whoa." Then, the rope would be tripped, and the hay would drop into the mow. Later, our Jeep was used to pull the rope, and we tried to ride along, often seeking to sit atop the back endgate for the ride. The photograph shows brother Ed at the wheel of the Jeep, with Shirley and Queenie the dog sitting in the front. I was sitting on the back tailgate. (As soon as this photograph was taken, Ed let out the clutch and I flipped over backwards onto the ground.) Instead of the Jeep, we often used one of our small tractors – sometimes the Oliver 55 and sometimes the Ford tractor – to pull the hay rope and I, even when I was as young as eight or nine, would drive both of these tractors. We all served our time making hay, lifting bales, stacking bales, and stacking bales in the barn, hay barn or pole shed. As the quote says, "Every person I know in the Midwest agrees that haying is hot, heavy, and dirty work" (Klinkenborg, *Making Hay* 140 (1986)).

Don, Larry, Merle, & Ed Wilberding

   Sal, Shirley, Queenie the dog, Ed and Merle

Merle: I remember they used a real short little sickle mower to cut the grass
           and the hay. Those work horses were probably fifteen or twenty
           years old by then. They probably had been there for a long time.

Evelyn: Yeah, I think Tony was probably one of the first ones to have a
           tractor out there for plowing and stuff like that in the neighborhood.

Merle: Is that right?

Evelyn: Uh huh. Everybody else had horses.

Merle: Used the horses all the time?

Evelyn: Yeah, yeah. But gradually all went to tractors.

Merle: I remember those horses pretty well. One was a big black one and
           the other one was sort of a, oh, a real light brown or whitish brown.
           Dick and Prince. That was their names.

Evelyn: Yeah, yeah.

Merle: Did you and Grandpa get electricity at the same time then? . . . .

Evelyn: Grandpa had a Delco. When they had in their house.

Merle: What's a Delco?[43]

---

[43]This term had special interest to me. The terms "Delco" and "Frigidaire" were often
used as generic names on the farm, even though they were intended as specific brand products.
We always referred to the refrigerator as the "Frigidaire." Later, when I moved to Dayton, Ohio,
I learned that the Frigidaire refrigerators were made in Dayton, Ohio, and that "Delco" was
derived from the Dayton Engineering Laboratories Company, which was incorporated in 1909
by Charles "Boss" Kettering and Colonel Edward Deeds and became the think tank for a
number of inventions for General Motors and its automobile production, including the self-
starting ignition system. Colonel Deeds and Boss Ket were major reasons for the development
of Dayton, and their legacies continue to benefit the Dayton community (Young, *Boss Ket* 85
(1961); Marcosson, *Colonel Deeds* 123 (1947)).

Evelyn: Well, a Delco was just a thing in your own basement, you know. That's what a lot of them had.

Merle: Like a generator?

Evelyn: Like a generator and then he had lights with about one plug-in in each room and lights. When they built that house they put that in.

Merle: They put a Delco in the basement?

Evelyn: Yeah, and they had electricity that way. I don't think they used it much for anything else. It was just mostly the lights.

Merle: That would have been before the REA then.[44]

Evelyn: Oh, yeah.

Merle: If you wanted electricity, that's what you had to do.

Evelyn: Yeah, yeah. And we didn't get the REA until Larry was a baby because during that time we got a refrigerator. After the lights went on.

Merle: That's the refrigerator you had for 40 years.

Evelyn: No, this one isn't.

---

[44]The term "REA" stands for Rural Electrification Administration, which was created by legislation enacted by Congress in 1936 to create a structure to provide electricity to the rural communities. Prior to this legislation, fewer than one farm in ten had electricity (Hamilton, *In No Time at All* 16 (1974). The rural electrification was typically a co-op effort in which each farmer had to contribute a small amount of money so that the local co-operative – in our case, the Glidden REA – would qualify for the federal loan to finance the installation of poles and lines to extend electricity to each farm. Indeed, it has been suggested that rural electrification did more to change rural America than any other single product of technology (*In No Time at All* at 18).

Merle: No, but the white one that used to be here.

Evelyn: Yeah, that's down in the basement.

Merle: Yeah, that's the one you got in 1941, isn't it?

Evelyn: Yeah, yeah.

Merle: Sure.

Evelyn: Yeah, that was the only one we ever had.

Merle: Yeah, that was a Frigidaire.

Evelyn: Yeah.

Merle: Because we used to always call it the Frigidaire.

Evelyn: Sure. Oh, we were thrilled to get that and the lights in the kitchen. You know, otherwise you had kerosene lamps. You'd have to clean the lamps and we had one gas lamp that you'd have to pump up in the kitchen which made a brighter light.

Merle: How did you keep the house warm? Those are big houses.

Evelyn: Well, there was a furnace and we had a lot of wood. They'd saw wood. Grandma would be out there too, sawing wood and . . . .

Merle: But there was a furnace in the house so that you just burned the wood . . . .

Evelyn: Yeah, yeah.

Merle: But, of course, on that East Place the furnace didn't go through the house because they had those registers in the floor and the ceiling, so that the only way heat got up to the upstairs was through those vents in the ceiling or from the floor.

Evelyn: Yeah, uh huh.

Merle: That's how the East Place was. I remember . . . .

Evelyn: Yeah, but when they built that West Place[45] then they had all the hot water was heated like through the, oh, it was right beside Grandma's kitchen stove, and it was hooked up to there, that it would heat water to go to the bathroom, like in the tub and stuff.

Merle: Say that again. What was that?

Evelyn: Well,

Merle: On the West Place?

Evelyn: On the West Place, yeah. When that was built new, they put . . . , oh, they put this water heater down in the basement, you know, in that front room where we had the milk separator, but they also had this tank right beside her cook stove which you always used. And that tank was hot all the time, you know. And that's . . . .

---

[45]Besides the woodburning cookstove (which also burned cobs, and we had a huge "cob room" in the basement to store the corn cobs left over from shelling corn for use in heating the house) in the kitchen on the West Place, there was a coal-burning furnace in the basement that heated the entire house. There was a separate coal room which would hold a truckload of coal which, hopefully, would last the winter. One of the chores that was handed down to the youngest person – and I had several years experience, beginning at about the age of six – was filling the "stoker" every night with coal by shoveling coal from the coal room. The stoker itself was a red bin with a lid which closed tightly because of the handle which hit under the latch. The stoker had a power auger which would periodically auger the coal into the furnace. After the stoker was filled with coal, the clinkers would have to be grabbed by the long iron tongs, removed from the furnace, left to cool, and eventually be removed from the basement. I was as happy as anyone when we converted the coal-burning furnace to an oil-burning furnace.

52

Merle: That kept and got the water hot.

Evelyn: Yeah, that kept the water hot, and it was hooked up just probably
          to the bathroom upstairs above that.

Merle: And was that your only source of hot water? No water heater?

Evelyn: No, that was the only way we got hot water.

Merle: Well, let's see. In those days you didn't have any kids in school.
          The kids went to country school close to the farm then. But what
          about high school? But there wasn't a high school then, was there?
          I wonder how long the kids went to the country school. You
          know, did they go to school for their entire first eight years down at
          the corner?[46]

Evelyn: Not the whole eight years.

Merle: Didn't they?

Evelyn: No.

Merle: How many?

Evelyn: They'd go until about until they'd have to make their communion
          and stuff. You know, you'd go, well, we sent Alice down there first

---

[46]The "school down at the corner" was one of the "one room schoolhouses that dotted
Iowa" (Hamilton, *In No Time at All* 133 (1974) – and, indeed, most of the Midwest – and
served as the cornerstone of education during our early years on the farm. This one-room school
house, known formally as Wheatland # 7 School, was located at the northeast corner of Section
No. 35 in Wheatland Township, in Carroll County, Iowa. The school house was on the
southwest quadrant of the intersection of the county roads that separated our farms. Now that
intersection is known as Falcon Avenue and 150th Street. This school house and virtually all
school houses in Iowa were white, not red. (Later, in the early 1950s, this school house was
loaded on a trailer and moved to a lot in Breda, Iowa, where it is now located at 407 Main
Street.)

and mostly because Gertrude[47] was teaching then. And you had to have so many kids to open the school.

Merle:	To keep the school open?

Evelyn:	Yeah. That's how Eddie got started so young. He started at four years.

Merle:	Because they needed him to get the school open.

Evelyn:	Yeah, I don't think they just needed Alice but then anyway she went, but then she didn't want to walk alone. Evie had to go along with

---

[47]This reference is, of course, to Gertrude Wessling, who was the teacher at the one-room school down at the corner to which Mom had earlier referred. In a separate taped conversation, Gertrude fondly remembered the country schools (as they were typically called) because everyone got individual attention and encouragement "and, well, afterwards, you got acquainted and got along." (*Wessling Tape* at 20). She recalled the shock from going from the country school to the Sisters' school and had these memories of her own childhood in which she tried to adjust to the rigorous education administered by the nuns:

> Hey, you just didn't know where you were at, and I was crazy about school. I thought I was a big shot, you know, at school, and when we'd go to the Sisters' school . . . . I didn't know a soul in that building. Boy, it wasn't very long: "Well, well, what kind of bandies do we have here? You have to have a special invitation to come out to class, do you?" And so on. Boy, I tell you, I was ready to bawl. I didn't know where in the dickens you belonged. And you stood there in a long line the length of those rooms, and you had to keep your feet parked that they didn't go over the crack of the floor, and if you stepped out of line someone tattled on you. I tell you, you talk about a reform school. . . . Still, I wanted to go . . . . we had slates yet, and I had one of these wooden slate pencils and it wore down. We were writing tables. Well, I knew my tables backwards and forwards, but my pencil was done for and I didn't know what to do. How are you going to sharpen it? And I started to bawl, and finally that sister came down there and she says: "What was the matter?" and I just showed it to her. Well, this Julie Thelen, she gave me a pencil to write with and I could go on again, but I was scared to say my soul was my own. I tell you it was really an awakening. (*Wessling Tape* at 10).

her. That's true. You can ask either one of those girls. They know that, that Evie had to start to school a year earlier because Alice didn't want to walk alone. So they started there.

Merle: And that was a mile down from the West Place? The school house was about a mile.

Evelyn: Uh huh, yeah. And then they had to go to Breda, which Dad had to take in then every morning.[48] And then Don probably started in Breda, because the rest of them was going, too. I don't remember him going to the country school.

Merle: Well, Eddy did.

Evelyn: Yeah, Eddy did because Gertrude came over that time. She couldn't get the school. There wasn't enough kids in that neighborhood to open the school. They was either going to the Catholic schools or there wasn't any in that neighborhood no more. So she thought if we could just sent Eddy, they could open the school. So we started Eddy at four.

Merle: Is that right?

Evelyn: Yeah. That's right.

Merle: And Gert taught everybody that was there. How many kids would have been in the school then?

Evelyn: Oh, about five or six. They had to have that many to keep open and everything . . . .

---

[48]There were certainly ten or twelve years when Dad had to take us kids every morning and bring us home every night. And that daily routine is probably well remembered by all of us, as we struggled to get everyone in the car in the morning to get the school and then waited every afternoon – generally from one to three hours – for Dad to pick us home and bring us home to do our daily chores.

Merle: In that the one-room school house?

Evelyn: One-room school house which was part of Kenny Snyder's house there.

Merle: Yeah, yeah, I know which house it is, yeah. Yeah, I remember when that school house was out there. I remember when they moved it.[49]

Evelyn: Do you?

Merle: Sure. Yeah, it was right on the corner on that . . . .

Evelyn: Anyway, I never dreamt that they would put Ed ahead in the second grade already. But see he graduated from, I think, was he sixteen or seventeen. He was the youngest one to graduate from the high school.

Merle: Yeah, yeah, I remember that.

Evelyn: I think he was four and he was sixteen, I believe, or something when he . . . .

Merle: He would have been seventeen when he got out of high school.

Evelyn: Seventeen?

Merle: He was born in '36 and graduated in '53.

------

[49]The old Wessling country school house (Wheatland # 7) had a long history in the community. As early as February 2, 1911, *The Breda News* reported that a "basket social was held at the Wessling school house, one mile west and three miles south of town. Quite a number of young folks went out from town and report a fine time." During their existence, these country schools served as educational, cultural, and social centers for the surrounding farms. (There is an extensive discussion of country schools in the *Breda Centennial Book*, Pages 228-231.)

Evelyn: Oh, yeah. And he would have been seventeen just that time. But anyway, he was pretty young.

Merle: He would have just turned seventeen.

Evelyn: Yeah, yeah. That's right. But that's how he got started – normally they didn't; now they send them to a pre-school at four. There's a lot of them around here. They just go a couple days a week, though. They've got a school out there by that, over where Schroeder[50] used to have that there old hilly place out there. They built a school out there.

Merle: Did they really?

Evelyn: Yeah. I've never seen it. But they say they got a school out there, and then there's some people living there and there's new houses out there, too. And, they call it a pre-school. They take kids out and they have to pay to go out there.

Merle: Okay, when you got married, Grandpa already had the West Place then, didn't he?

Evelyn: Oh, yeah.

Merle: He bought that from the Wesslings.[51]

Evelyn: Yeah. Yeah, that was where Grandma lived.

---

[50]This is a reference to Maggie Schroeder, a long time neighbor, and that "old hilly place out there" is a reference to what we always called the "Hazel Nuts" which was then a series of hills and ravines with steep grades on the gravel roads, making them dangerous to drive on. These hills were located one and one-half mile east of our West Place. These hills have since been regraded and there is now a small public roadside park overlooking the ravine.

[51]See the discussion on the acquisitions of the farms in Footnote # 1.

Merle: Yeah. Do you remember anything about when he bought the East
          Place? Or that would have been before you were married, too.
          Yeah, apparently there was some argument about it. That was
          Anton's place. I guess there was some disagreement as to whether
          Grandpa was able to buy it or whether the other kids got it. Alice
          sent me a court record which said that Grandpa filed a lawsuit
          against the estate for the right to buy it.[52]

Evelyn: Oh, is that right?

Merle: That he had some agreement with his Dad, Anton, to buy it and they
          upheld it and he bought it.

Evelyn: Oh, he did?

Merle: Yeah. For some amount of money. And I suspect it may have been
          that he just wanted it, as opposed to going to an auction, you know,
          because it was right next to the other place he had.

Evelyn: Oh, yeah.

Merle: Yeah, Alice sent me those records. They were interesting.

---

[52]The East Place was bought by Grandpa Frank Wilberding as a result of a law suit he
filed against the Estate of Anton Wilberding, in which he sought to enforce an oral agreement
which had been made and reduced to writing, but had not been signed, giving Grandpa
Wilberding the right to purchase the East Place for $ 20,000 from Anton Wilberding and Maria
Wilberding after their deaths. This Decree was signed by Judge F.M. Powers on March 7,
1913, shortly after the death of Anton Wilberding in 1912 (Book 16, Page 473, Deed Records
of Carroll County, Iowa). However, the Order of Conveyance of Real Estate was not issued
until April 19, 1926, shortly after the death of his mother, Maria Wilberding. This was
consistent with the written (but unsigned) terms of the agreement, which reflected that Anton
and Maria wanted to own the farm until both of them were dead. Although there are no records
to prove or disprove this theory, I suspect very much that this was a "friendly law suit" in that
all the parties in interest agreed to this result, but that the parties needed a court order to validate
the chain of title for the real estate.

Evelyn: Oh, I didn't know that. She didn't . . . .

Merle: Did the Jaspers ever visit, like that Mrs. Jasper was Grandpa's sister
          and she apparently died way before you ever met Dad.[53]

Evelyn: Yeah.

Merle: Are they the ones who were in Nebraska that . . . ?

Evelyn: Well, there are Jaspers out there, but there also was a Jasper that
          was married to Tony Pullman. And she's dead now too, but that's
          where the Pollmans come from. They're partly our relation too.

Merle: Because they married the Jaspers and the Jaspers had married a
          Wilberding? Is that it?

Evelyn: Well, she was a Jasper and he was a Pollman. I know they always
          called themselves relation. They still do really, like Walt Pollman
          and. . . .

Merle: Do they really?

Evelyn: He lives on that corner on the east side just across from . . . .

---

[53] The Wilberding-Jasper relationship is very interesting, but because we had little
contact with them growing up, the relationship is also somewhat confusing. Grandpa Frank
Wilberding had a sister, Agnes Wilberding, who married John Jasper. Agnes and John Jasper
lived in Elgin, Nebraska. Among their children were Mary Jasper and Frances Jasper who
married, respectively, Charles Wilberding and Herman Wilberding, both of whom were sons
of Anton Wilberding of Remsen, Iowa. This Anton Wilberding would have been a first cousin
to Grandpa Frank Wilberding. These marriages were between two sisters of one branch of the
Wilberding family and two brothers of a different branch of the Wilberding family. (Much of
this information came from Lawrence and Anne Wilberding of Meriden, Iowa, to whom I am
grateful. Lawrence Wilberding is the son of the Charles Wilberding who had married Mary
Jasper.)

Merle: You never remember that Maria Wilberding?[54]

Evelyn: No . . . .

Merle: She died before you ever met Dad.

Evelyn: Yeah. Yeah.

Merle: And you met Anton's mother, . . . er Anton's wife, Maria? She died
          in '26 or is that when you just went to the funeral?

Evelyn: I just went to the funeral there.

Merle: That would be Grandpa's mom.

Evelyn: Yeah. Yeah, because Mrs. Wilson said I could go to the funeral and
          I really never knew her and I don't think she came to church any
          more. She was getting up there in years. And, she just thought I
          would like to go to church, but I never did see the lady.

Merle: Who was Mrs. Wilson?

Evelyn: Oh, she was, well, you know, where they always had that drug
          store. Do you remember that Lottie Jones . . . . [55]

Merle: Oh, sure. I was in there many times when I was a kid.

------

[54]Maria Wilberding was the daughter of Anton Wilberding and Maria Grever and was
a brother to Grandpa Frank Wilberding. She was born August 20, 1870, and died July 17,
1918.

[55]The Jones Drug Store was established by Dr. U.C. Jones, in 1881. His daughter,
Lottie Jones, operated Lottie's from 1915 until 1964. Then, its entire interior – including the
shelving, apothecary scales, and her famous candy cases – was purchased and removed to
Chicago where her store was recreated in Chicago's Hall of Fame on North Shore Drive
(*Breda Centennial Book* at Page 305).

60

Mrs. Wilson's Dry Goods Store, Breda, Iowa

Evelyn: Yeah, well, Lottie Jones' sister was Mrs. Wilson. She married a guy named Wilson and then they had Hazel and what was that other one's name. Hazel and they had another girl. She had those, and then, I don't know whether her husband died or what, but anyway, she run that store that Wilson's, Mrs. Wilson's store. She had a dry-goods store then. That's who I was working for.[56]

---

[56]Mrs. Wilson's store was on Main Street in Breda, Iowa, and advertised itself in *The Breda News* as "The Cash Bargain Store," Mrs. I. E. Wilson, Proprietor. Ida Wilson sold all kinds of dry goods, including hosiery, blankets, shirts, fabrics, and specialty ladies wear. In addition, she carried a variety of chinaware, household goods, and giftwares.

Merle: And it was in that place where Lottie's place was. Is that where Mrs. Wilson . . . .

Evelyn: No. No, Lottie was on the west side and Mrs. Wilson's was up there about where the – Tiefenthalers probably got that place now.

Merle: And what did Mrs. Wilson sell?

Evelyn: Oh, she had dry goods where you'd measure clothes by the yard and calico. You could buy pretty prints when that came in, for a quarter a yard, you know, and you'd have to measure that off. And then she'd take in eggs, too. And you have to candle each one of those eggs. They'd bring in a box of eggs for such an egg case, and I'd have to candle them. She had . . . . You'd have to look at every egg to see if there was a chicken in it or what.

Merle: Is that right?

Evelyn: Yeah. We always had to do that, and then they'd get paid for that. You know, people didn't always have cash. And then she'd take them to the creamery there.

Merle: Yeah, in fact I saw an ad in one of those old *Breda News*. I don't know what store it was. Eggs taken same as cash.

Evelyn: Yeah, well Determans always took eggs in, too.[57]

---

[57]Mom, of course, is talking about when she was working for a store which was buying eggs, and then recalling that she could pay for food at Determan's Grocery Store with eggs. This was confirmed in the "Nostalgic Memories" that "people brought their eggs to Determans. where the eggs were candled and valued. If there was more egg money than groceries, they got a credit." (*Breda Centennial Book*, at Page 387). If the egg money was not sufficient, then the grocery store would carry it as a receivable until springtime when hens were laying again. That practice had stopped by the late 1940s. But we still gathered eggs every night, usually after dark – and no one who has had to shoo old hens off their nests in the dark so that he could pick up the eggs could ever forget that experience – and bring them back to the house, down to the basement, and then pack them in the double egg crates for sale to the Breda Creamery.

Merle: Is that right?

Evelyn: Yeah   Determan's store used to take in eggs in the old days, but later, we just would sell the eggs and cream to the Breda Creamery.

Merle: On a different topic, did you and Dad ever take that train anywhere?

Evelyn: Oh, let's see. I don't think so. I don't know. The places I remember, we always took cars.

Merle: Yeah? I was just curious. I know they had all those passenger trains then, and I just wanted to know because I was looking in *The Breda News* and it had schedules for some of the trains, you know, the Chicago, North Western . . . .

Evelyn: No, I don't think we ever took a train. Even when we went to Joliet. We was out there for a funeral. Clarence drove that time. Clarence, Clara, and Tony and I went. Do you remember that old John Agnich?[58]

Merle: No, I don't think so.

Evelyn: You may remember them as some of the Goecke relation that used to come out to the farm every so often to visit us and the other Goecke relatives.

Merle: From Joliet?

---

[58]Although I immediately said "No" when Mom asked me if I remembered "that old John Agnich," I did later recall his rare visits to Iowa from his home in Joliet. Although I would have been very young at the time, perhaps five to seven years old, I do remember him as being part of the "Goecke relation" and that he was a guard in the Joliet prison. From a relationship connection, it appears from death cards and obits that the wife of John Agnich was a daughter of Elizabeth (Boecker) Skoff (1878-1963) and that Elizabeth Skoff was, in turn, a sister to Grandma Margaret (Boecker) Goecke. In short, this meant that Mom was a first cousin to the wife of John Agnich.

Evelyn: Yeah, he's dead now, too. And Lawrence and Jewel are both dead and I don't know how many, Johnny is dead out there and . . . .

Merle: When did, when Grandpa moved to Carroll was there a reason why he moved to Carroll as opposed to Breda? Was he, like you always say, only willing to use the Carroll banks and move to Carroll? I just wondered if he . . . .

Evelyn: It just seems to me that Grandma wanted to go to Carroll. She didn't want to live in Breda.

Merle: Really?

Evelyn: Yeah. I think so.

Merle: Huh.

Evelyn: I don't know for sure what the reason was but . . . .

Merle: So they moved to Carroll about the time, shortly after Larry was born, didn't they? Wasn't Larry born and you were on the East Place?

Evelyn: On the East Place, yeah.

Merle: And then you moved to the West Place before I was born. I was born in '44.

Evelyn: Yeah. Uh-huh. Yeah, you and Shirley were born . . . .

Merle: And I was on the West Place then or you were on the, we all were on the West Place then.

Evelyn: Yeah. And they moved that time and we moved over to the West Place.

Merle: Well, Grandpa would have been pretty old then already.

Evelyn: Yeah, he was.

Merle: He was born in '77 so that's 23 and 43 is 66. So he would have been 65, 66 years old when he moved off the farm.

Evelyn: Uh-hum. Yeah.

Merle: I was expecting he was a lot younger, but he wasn't. He was, worked a hard life by then.

Evelyn: Yeah, yeah. Yeah, I remember, too, when they were debating what house to buy and Grandpa was out looking for one and he found this nice brick house.[59]

Merle: Do you remember them moving?

Evelyn: Oh, yeah. Un-huh.

Merle: When you moved from the East Place to the West Place, how did you do it?

Evelyn: That was something because, you know, when they was going to move to Carroll we wanted the guys that was going to haul them to Carroll to come over, which they did, and it was cold. It was about twenty below zero that day.

Merle: Is that when they moved to Carroll?

Evelyn: That's the truth. When they moved to Carroll. But they did stop into our place because we had a piano at that time and the stove. And Dad asked them if they'd stop and just pick up those two big

---

[59]Grandpa Frank Wilberding's house was at 1215 N. Adams Street, Carroll, Iowa.

things and take them over to Grandpa's before they would load up their stuff. So that's what they did. But, Lord, that was a cold, cold morning.

Merle: What did they have, a truck? Where did Grandpa Goecke live when you were born? Did you live out on that Goecke place that was north of the blacktop out here?

Evelyn: No. I was born on a farm east of Carroll.

Merle: East of Carroll?

Evelyn: East of Carroll. Dad just rented that.

Merle: Is that right?

Evelyn: Um-huh. He rented that and see his parents, I think they homesteaded both of those, the Goecke, the old Goeckes, I think they homesteaded . . . .

Merle: Frank Goecke, Grandpa's Goecke's dad.

Evelyn: Yeah. And the one, well, Joe Goecke got, and then Dad eventually bought the one on ours when they had a sale then.

Merle: Yeah, but that was north of Mt. Carmel.

Evelyn: Yeah, that was the one north of Mt. Carmel. But I was born east, where we just rented. And then we lived there probably a couple years or so, and for some reason or other, either Dad wanted to move, or had to move, maybe, or something. And then we went to Maple River . . . . That was where we was real close to where the O'Briens lived.

Merle: Is that right?

Evelyn: Yeah, we used to go sleigh riding, I don't just remember Bing, but then there was Frank O'Brien about my age, and he used to have a little sled. He's dead now. And we'd go sleigh riding, but it was close to Maple River and it was just a poor house. They've got a new house on that place now. But we, and then that's when they was going to sell the two Goecke places.

Merle: Add a little bit more, and then I'm going to go up in the attic, and see if there's anything up there.

Evelyn: I was going to say, what I think is up in the attic and I don't know how good it is anymore, was such a big, heavy lap robe. That's the kind we used to use in our sleds and stuff.[60]

Merle: Yeah. I was . . . .

Evelyn: I don't think so.

Merle: There may not be.

Evelyn: Sal wants the trunk that's down in the basement.

Merle: Oh, no. I'm talking about the contents.

Evelyn: Oh.

Merle: If she wants the trunk, she should have the trunk.

Evelyn: Yeah, she always said that and I said she could have that trunk. I made a note.

Merle: That's a nice trunk. That's fine.

---

[60]This heavy lap robe was about four feet square and had a coarse dark brown cloth on one side and a quilt-type covering on the other side, and was very heavy around the sides, almost like it had weights in it.

Evelyn: She asked me that different years ago. I used to ask the kids what they'd like and I do have a list in my cedar chest. But you're getting the cedar chest.

Merle: The cedar chest?

Evelyn: The cedar chest is yours.[61]

Merle: My absolute first choice.

Evelyn: I know, that's what you said. So I've got that down. And they say that, that's as good to me. You don't have to have it in the whole will.

Merle: Oh, no.

Evelyn: I'm trying to find something that each one would like. Shirley gets the organ.

Merle: Yeah, yeah. Yeah, she's the only one that would use it.

Evelyn: Yeah.

Merle: She's the only one that would use it.

Evelyn: Yeah, she probably wouldn't have time to use it anymore.

Merle: Well, how old were you when he bought that Goecke farm. He bought that one.

Evelyn: Yeah.

---

[61] The cedar chest was always used by Mom to store her special mementos and the special family records. Despite its cedar lining, she never used it to store clothing. In many ways, it served as the family safe.

Merle:	How old were you then?

Evelyn:	Well, I could have been about, I don't know how soon we moved off, but I graduated from the eighth grade in Mt. Carmel and that was the last year I went. Now I don't remember if he bought that farm a year or two before that.

Merle:	Okay. When you went to school in Mt. Carmel you were living on that Goecke farm then and went to Mt. Carmel.

Evelyn:	So, I was born and I made two grades in one year. They pushed me ahead.

Merle:	Uh-huh.

Evelyn:	So I would have been about thirteen or fourteen.

Merle:	Did you go to school anywhere besides Mt. Carmel?

Evelyn:	No.

Merle:	Did you go in Maple River before that or when you started?

Evelyn:	I was too young to go when I was in Maple River.

Merle:	Okay, so when you started school, you started at Mt. Carmel?

Evelyn:	Well, I started, we had country schools too, east of our place, where those Goecke places was, and in the winter time this was the only schooling that I got. In the winter time the snow was so bad that we couldn't get to Mt. Carmel excepting the year when I was going to go communion and . . . .

Merle:	Because of the snow, you mean?

Evelyn: Because of the snow, we just had horses. We had a blind horse that us kids . . . .

Merle: Flory.

Evelyn: Yeah, yeah, I believe that was it. How did you know?

Merle: Well, you told me before. You were talking about, when we were talking about the hot bricks and . . . .

Evelyn: Yeah, yeah . . . the blind horse, yeah. And anyways, we would stop about Christmas time or whenever the snow was getting bad, and then we'd go to the country school. They had teachers there, and there was, oh, there was quite a few kids. There was a lot of them that couldn't get to, you know they didn't have cars, you couldn't drive cars in those days because there was no way to keep them. They'd freeze up.

Merle: They'd freeze up.

Evelyn: Yeah, they didn't know about the stuff now. So, we'd go there and then that one year I got put ahead because I was so smart. Annie Snyder told me that, just not too long ago. We was talking about that, and she said, I said wasn't you in my class? She said, "Well for a while, but you know you got put ahead that one year." And that's the way I remember it, too.

Merle: Really?

Evelyn: Yeah.

Merle: Okay, so you started going to country school for several years and then you got pushed ahead.

Evelyn: Yeah, so I really only went seven years to school.

Merle: At Mt. Carmel, then?

Evelyn: Yeah, yeah.[62]

Merle: How would you get to school in Mt. Carmel then? Would you go?

Evelyn: We had a blind horse. That's the one we'd take. And then that one
           winter when I had to go to communion, but you had to be there for
           the instructions the whole year. [63] And then Lizzy was, she was in
           the upper grades above, probably eleventh or twelveth grade, and
           she got to be a teacher, you know. She got smart. And anyway,
           that year, when Lizzy wanted to go then and I was to go to
           communion, so we had to board with the nuns.[64]

Merle: Did you really?

Evelyn: Yeah. And we slept up on the third floor there, of that Mt. Carmel.
           The same building that's there.

---

[62]The school and convent in the village of Mt. Carmel (Carroll County, Iowa) were
constructed in 1912, when Mom would have been eight years old. The *Quasquicentennial
Directory* ("*Mt. Carmel Directory*") for Our Lady of Mt. Carmel Church, published in 1994,
confirms Mom's recollection. "This school was built during an era when children left their
horses at a livery stable by day so they could attend classes. Later many boarded at school,
especially in the winter" (*Mt. Carmel Directory* at 4). Other parishioners echoed these
recollections: "Children walked, rode horses, or used horse and buggies to get to school. A
barn stood near where the rectory is today. Students tied up horses during school. They took
turns providing hay and oats" (*Mt. Carmel Directory* at 10).

[63]Although it was badly torn, Mom had a very fancy religious certificate attesting to her
First Communion on June 28, 1917, at Mt. Carmel, and her Confirmation on October 12, 1919,
at Mt. Carmel. This certificate was signed by Reverend John B. Baumler, Pastor.

[64]It was amusing for Mom to talk about boarding with the nuns, both in its own context
and because of the many times it seemed like we were boarding with the nuns after we ended
up eating dinner with them because Dad was hours late in picking us up after school. I think
we all remember eating at the convent in the dark, only to eventually get home to do our chores.

Merle:	Of the convent in Mt. Carmel?

Evelyn:	No. It's the school, but the sisters had one part of it where there was always a lot of sisters. Now they're using part of the place because I've been in Mt. Carmel several weeks ago, and . . . .

Merle:	So you lived in the school for a while, because the weather was so bad.

Evelyn:	Yeah, on the third floor, and there was one side of them that was for boys and then the girl's room was over here, and the sisters always had one little cubby hole where they'd stay so you couldn't talk, you know. You had to shut up.

Merle:	Who drove the horse to school? Who was in charge of the reins?

Evelyn:	Well.

Merle:	You? You?

Evelyn:	Oh, I don't think I ever did. There was always older ones than me and then Clarence.

Merle:	Like Frank, was he?

Evelyn:	I don't remember Frank going to school with us. I mean just the country school he might of went to, but then we lived in a few places before Frank got here. I don't think he ever graduated.

Merle:	Frank was a lot older. Clarence was the youngest one.

Evelyn:	Clarence was the youngest one.

Merle:	Yeah. And Frank was like the oldest?

Evelyn:	Uh huh.

Merle: What did the horse do during the school? During the day when you
          were at school.

Evelyn: They had a little barn always in back of that one building there, and
          everybody, some had mules that they dragged in, I know the Grotes
          always had a white mule.

Merle: Is that right?

Evelyn: Yeah. And . . . .

Merle: So you'd go to school and tie the horse up in the barn?

Evelyn: Well, you'd unhitch them and take them into this shed, and then
          you'd have to go down at noon and bring them some corn and put
          corn in the box and take your dinner and then you'd get home again.

Merle: No kidding?

Evelyn: Uh uh. And there was an awful lot of buggies and stuff that the kids
          would, there was no cars or anything like that.

Merle: Yeah, yeah. I'll be darn. Didn't you, when you were living on that
          farm, didn't you tell me once that there were gypsies that would
          come through from time to time?

Evelyn: Yeah. There was a lot of gypsies or Indians whichever you'd call
          them. They looked like Indians, and they was always carrying a
          shawl around them like that. And then they'd be begging for food,
          you know. And they'd have . . . .

Merle: Did they come to the farm?

Evelyn: Yeah. They'd usually, they wouldn't turn in. They'd park on
          corners, you know, like we was kind of in between two miles, and
          it was either the east one or the west one where they'd park and

they'd stay there overnight, and they always had a whole bunch of horses that they'd lead along with them. And then they'd go out begging.

Merle: Is that right?

Evelyn: Yeah. And they'd come up and, oh, my Dad was always tough. He never was scared of them. If they'd ask for anything we'd give them a loaf of bread or something like that. But not Dad. Dad said, "Nope. No way. You go on your own way."

Merle: Is that right?

Evelyn: He was tough. And then that one Indian says, or gypsy, whatever she was "Government gives us a right to go over any place. We can go any place we want to," he say. And Dad says, "The government don't own this place. This is mine and you get out of here."

Merle: Is that right?

Evelyn: Yeah. That's just how he told them, and us kids were so scared that they'd be mad and kill us all through the night, you know. We was so scared, but Dad was tough. He wasn't scared of them.

Merle: I remember when I was about eight or nine years old ,gypsies coming across the west road there. God, I was probably eight, nine, ten years old. And they came by horse and wagon.[65]

Evelyn: Oh, yeah. Yeah, they probably did that.

Merle: This would have been in the early 50s.

---

[65]This group and their horse and wagon passed by our West Place without incident, but I still remember everyone in the family huddled at the corner of the house to watch them pass by. In the paper the next day, there was a report that Sheriff Al Thorpe had been called and had escorted them out of Carroll County.

Evelyn: There was a lot of them and then there was horse traders, too, you know. They'd have a whole bunch of horses and then they'd have, well, just a wagon, you know, like those things . . . .

Merle: Yeah. Would they stop and like to try to trade horses with your Dad?

Evelyn: I don't remember if they ever tried to trade, but I think they probably would do that along the way if they'd find somebody. But they had all those pretty horses, really though, just like you'd see in Montana, probably. Those spotted ones and stuff.

Merle: Did the gypsies steal a lot of food? Chickens and things like that?

Evelyn: They said yeah. I remember on the one farm where the Olerichs lived there. She went up to the house and the women gave her a loaf of bread. And then she watched her as she left, and this gypsy grabbed a hold of a chicken and they used to carry them, they had these big aprons on, you know, and they put the bread in there. She grabbed a live chicken and put it right in the basket there or her lap wherever she had and stuck that. But they didn't say anything. They thought just let her take one chicken so she get out. Here comes some of those crazy jets.[66]

Merle: Yeah. Was your Dad born in the United States?

Evelyn: Yeah.

Merle: He was born in the United States. But his dad was born in Germany, Frank Goecke?

Evelyn: Frank? I suppose, I don't know if I've got that down in the Bible.

------

[66]This last sentence, of course, has no relation to the rest of the conversation. Mom was just commenting on her annoyance with the low flying military jets which often interrupted her.

Merle: Did they ever talk about that? You know, like Grandpa Wilberding always did. Did the Goeckes ever talk about it at all?

Evelyn: No. I tell you, the last I remember and probably the only time we didn't always, Grandpa Goecke stayed with the Joe Goeckes there on that farm and all I can remember is that he always, he had a great big beard and . . . .

Merle: Frank Goecke?

Evelyn: Frank Goecke. Yes, my Dad's father. And I don't remember hearing them saying much of anything. We never, Grandma was quite lively. She was a small lady but she lived to be . . . .

Merle: That's Gertrude? Gertrude Lauman?

Evelyn: Gertrude Lauman, yeah. But she was kind of a tiny little lady, but she liked to dress up pretty though, I remember.

Merle: Really?

Evelyn: Yeah. And she wore white shoes which we thought was pretty classy at that time, and she was old. But she liked pretty things.

Merle: They were both alive quite a while. Do you remember when they died?

Evelyn: Oh, I think I have that in my Bible. Bring my Bible.

Merle: Yeah, I've got it. Let's see.

Evelyn: No, I don't remember.

Merle: Gertrude Lauman died in 1915.

Evelyn: Oh.

Merle: And Frank, Grandpa died in 1909.[67] So you would have been real young.

Evelyn: That's what I thought. I was just a pretty little girl.[68] Not pretty – pretty little.

Merle: Pretty little?

Evelyn: Yeah.

Merle: Yeah, you would have been five years old then and only ten years old when, eleven years old when Gertrude Lauman died.

Evelyn: Yeah, yeah.

Merle: And, I think I remember that your grandpa, Frank Goecke, had owned some farm land?

Evelyn: Yeah, I think it was after Grandma died that they had the sale of the two farms. Like I said, I was just a little girl then and I don't remember for sure. But, I think the Goeckes both homesteaded those two farm places really, I do. You'd have to look it up in the courthouse probably. But then I think that's what they used to say that . . . .

---

[67]Frank Goecke was born on January 6, 1830; he married Gertrude Laumann (sometimes spelled "Lamann") in the 1860-1865 period at Hazel Green, Wisconsin, and he died on January 11, 1909. Gertrude Lauman was born on February 17, 1834, and died on February 12, 1915. One of their children was Grandpa William Goecke.

[68]This exchange between us was classic "Mom." She was obviously trying to explain that she did not remember her Goecke grandparents very well because she was very young when they died. So, after she said that she was just a "pretty little girl" she caught herself with modest embarrassment, gave a little laugh, and then, almost shyly, explained that she wasn't saying that she was "pretty," but was only saying that she was "pretty little." Mom was always very modest about her own talents and virtues and she would not want anyone to ever think that she was claiming to be a beautiful child.

Merle: Yeah. Alice sent me a copy of Frank Goecke's, your Grandpa's will,
           and it was handwritten and he had signed it with an "X."[69]

Evelyn: Oh, is that so?

Merle: Uh huh.

Evelyn: Yeah, they probably couldn't read or write much in those days, you
           know, they were just poor farmers..

Merle: Yeah, I understand. So you lived on there, you graduated or went
           to Mt. Carmel through the eighth grade and . . . .

Evelyn: I graduated from the eighth grade and I had the measles at that time,
           and I didn't get to that either.

Merle: Is that right?

----

[69]The Last Will and Testament of Frank Goecke, Sr. was executed on November 20,
1901. He died on January 11, 1909. The interesting thing about his Last Will is that it was
handwritten, but not by Frank Goecke. It is evident that he was not able to read or write and,
consequently, he signed his Last Will with a mark. This is reflected in the following conclusion
of the Last Will:

In witness whereof I, Frank Goecke, Sr., have hereunto set my hand this 20th
day of November, 1901.

                              His
            Frank   **✗**   Goecke
                              Mark

Signed by the said Frank Goecke, Sr., as his last will, the same first having
been read over to him in our presence, with his mark, in the presence of us,
present at the same time, who, in his presence and at his request, and in the
presence of each other, have hereunto subscribed our names as witnesses.

            /s/   A.J. Olerich, Carroll, Iowa
            /s/   Frank Ludwig, Mt. Carmel, Iowa

Evelyn: Yes.

Merle: No kidding.

Evelyn: Went seven years and they had to put my diploma[70] on an empty
          chair, like a vacant chair, because I had got the measles and was so
          sick.

Merle: Is that right?

Evelyn: Yeah.

Merle: Oh boy.

Evelyn: I never had much fun in my life. Everything went haywire. I got my
          graduation picture around here some place, though.

Merle: Do you really?

Evelyn: Yeah. There was only four of us.

Merle: Really?

Evelyn: Carl Stork, he's living yet, and then me and Agatha Highland and
          Pauline Berger. Agatha I haven't heard from in so long. She used
          to write to me all the time at Christmas, but we kind of stopped that

---

[70]Although Mom attended the parish school at Mt. Carmel, her eighth grade diploma
(like all others, including Dad's) was actually issued by the Carroll County Superintendent of
Schools, George Galloway. Her diploma was dated May 31, 1918. This is consistent with an
1872 Iowa law creating the County Superintendent's position which included general
supervision of all schools in the county, and the authority to hire teachers and issue diplomas
(*Breda Centennial Book* at Page 227). Mom's diploma recited that she had completed the
course of study covering the work of eight years in the Common Branches. (Although Dad, of
course, went to a different school, his diploma is the exact same certificate as Mom's, including
being signed by George Galloway as the County Superintendent, but his certificate is dated
June 1, 1920.)

with getting old. So whether she's living or not. There was only four of us that graduated that year.

Merle: And you stayed on the farm after that?

Evelyn: Well, not just too long. In fact, it might have been right after Grandma Goecke died, that's when they sold the farms. And my Dad bought this one and Joe bought the other one, and then Dad got the notion he'd like to sell it, you know and he could, because the land was going up a little bit and he'd like to move to town. He had this hernia and he couldn't stand hard work no more.

Merle: Really?

Evelyn: Yeah, that's what he said.

Merle: He had a hernia?

Evelyn: Yeah, he had a hernia and he would never go to the doctor to get it; he didn't until just those last couple years when he got so bad he almost strangulated from that. And then he was in the Sioux City Hospital. That wasn't too many years before he died. He was quite old. And he was in Sioux City then. He just had to have something done but he was always scared to go for a surgery. Scared to go or didn't want to go. He thought if he didn't do anything it wouldn't bother him, but that why I guess he wanted to retire so young. But he was very young when he retired. Then he bought that place east of Breda. That twenty acres. Do you remember that?

Merle: No.

Evelyn: It's about twenty acres. The house was torn down along time ago, and it is all gone now.

Merle: But it's straight east of Breda?

Evelyn: Straight east of Breda and was on the, if you're coming from Breda,
          it would have been on the right-hand side.

Merle: What, out a couple miles?

Evelyn: No, it was on just about a half a mile.

Merle: Really?

Evelyn: Yeah. And we used to raise strawberries and he loved that. And he
          picked strawberries and we had a big patch. We all had to pick and
          then he'd take them into town and sell them for so much. He would
          make all his own little boxes. And I got the hammer, yes, which a
          lot of them probably won't know what a hammer like this was for.
          But, he ordered that stuff and this little hammer.[71]

Merle: That was your Dad's?

Evelyn: Yeah, he had little tiny nails.

Merle: Is that right?

Evelyn: Uh huh. And there was those boxes like you'd get strawberries in
          and he'd put them, they'd come flat. You'd have to fold them up and
          put just a little tack in each corner to hold those boxes together.
          And that was the hammer specially made to just handle that little
          stuff.

Merle: Is that right?

------

[71]At this point, Mom reached in her "junk drawer" and pulled out this small hammer,
which I would describe as a tack hammer and showed me that it still had the original weathered
handle. She gave it to me and I now have it framed with a note that it was made especially for
Grandpa Goecke when he was raising and selling strawberries.

Evelyn: Uh huh. And I've always kept that. And we'd all have to go out there, us kids that was at home yet at least, and pick those strawberries early in the morning. He made pretty good money on them because the strawberries was nice.

Merle: No kidding. I've never seen this before.

Evelyn: Yeah, probably not.

Merle: So you never lived out there. Your dad just bought the property out there, or did you . . . .

Evelyn: No, we lived out there. Yeah. Lizzy had her wedding out there and so did Teresy and I think Mamie did too yet.

Merle: Really?

Evelyn: That they was all married out there. Now I mean they lived out there and so did Frank and Bess. We had the wedding for Frank and Bess.[72]

Merle: Is that right?

Evelyn: Yeah. The Eilers for one thing, I think they was poor and they, I don't know, they just kind of disappeared. I don't know what happened to them. But anyway Bess was a nice little girl. Of course, one of them married that Patrick's mother, she was. The women that run the beauty shop. Yeah. Vanderheiden, that's who she married. Anyway, that . . . .

Merle: How long did you live on that twenty acres?

---

[72]Frank and Elizabeth ("Bessie") Goecke were my sponsors at Baptism. Frank Goecke (who was Mom's older brother) was born on February 12, 1898, married on April 4, 1921, and died on March 1, 1970.

Evelyn: Well, not too many years and that's when Dad wanted to build that house in town.

Merle: Okay, the one with the stone baskets in front.[73]

Evelyn: Yeah, that's right. That's when they started building that house and before . . . .

Merle: Your Dad built that.

Evelyn: Yeah, Dad built that. Him and his brother John Goecke helped. He used to build houses, too.

Merle: His brother, John?

Evelyn: His brother, John, yeah, that lived in Carroll. But he'd be up there and they built the house practically, mostly themselves, you know, and he plastered it, and I don't know if you seen a house before it was plastered, but they put strips of . . . .

Merle: Oh, those laths.

Evelyn: Those laths, yes. And then before they plastered.

Merle: Yeah I learned that when we tore down the wash house on the East Place, when we ripped it down.

Evelyn: Yeah, that's how they used to make them.

Merle: We had to rip out all those lathes in that old plaster, sure.

---

[73]The house with the stone baskets in located at 307 Bruning Street in Breda, Iowa. Grandpa Goecke built this house in 1925 ("70 Years Ago," *The Breda News,* March 29, 1995). Grandpa Goecke had built those baskets, just as he later built similar stone work at his cottage on Spirit Lake. There is a discussion of his cottage in Footnote # 79.

Evelyn: Yeah, I remember we went over there and we tacked those things on there before the plaster.

Merle: No. No. I remember tearing that down.

Evelyn: Yeah. But, before we got the house finished then, that people who lived right next to it, it was Dr., he was a veterinarian. And I can't think of his name, anyway he was living in that house next door to the right-hand side there, and so Dad decided we could rent that house then after he gave up, and we rented that and lived in that house until we had the new house done.

Merle: Is that right?

Evelyn: Yeah, yeah.

Merle: Okay. So when you were, how old were you then when you moved into that house?

Evelyn: Well, in the new house?

Merle: Eighteen, twenty?

Evelyn: Yeah. I was about eighteen or nineteen, something like that.

Merle: And Mrs. Anton lived across the street then? Anton Wilberding?[74]

---

[74]Great-Grandma Maria Grever Wilberding, the widow of Anton Wilberding, lived at 306 Bruning Street, right across the street from the house that Grandpa William Goecke had just completed building. Great Grandma Maria Grever Wilberding, the widow of Anton Wilberding. The county records show that Anton had purchased Lot # 7 (the house directly across the street) of Block 9 in the First Addition to the Town of Breda Plat from John & Mary Ricke for $ 900.00 (Book R, Page 461, Deed Records of Carroll County, Iowa). Anton Wilberding also purchased the adjoining lot to the east, Lot # 8, from Frank Bruning for $ 65.00 (Book R, Page 619, Deed Records of Carroll County, Iowa).

Anton & Maria Wilberding and their daughter, Maria

Evelyn: I think so. I think that they were both there. I didn't pay much attention to them.

Merle: Do you remember them?

Evelyn: Oh yeah, but in those days I was young, and we were busy, and the Wilberdings were elderly and pretty much stayed by themselves.

Merle: Was that how you met Dad? Because he lived over there?

Evelyn: No, I don't think that's how I met him. No, it wasn't because his family lived across the street in Breda from my folks.

Merle: Well, do you remember how you met Dad?

Evelyn: Oh, yeah, I do remember how I met him.   Marie Luchtel and I was very good friends, and Luchtels lived in that house where Gladys [Oswald] Nieland is living in there now. And it was just a mile, hardly a mile, from where the Wilberdings lived, you know. And then there was . . . .

Evelyn Goecke Wilberding

Merle: You mean the Luchtel farm out there.[75] Is that what you're talking about?

Evelyn: Well, yeah. And it was their folks that lived in that house, and then Marie and I was always real good friends.

Merle: So, then, you would visit Marie Luchtel at their farm.

Evelyn: Yeah, sometimes we would get together out at her farm.

Merle: So the Luchtels introduced you to Dad?

---

[75]The Luchtel farm was located one-half mile west of the West Place.

86

Evelyn: Yeah, Marie and Bertha and me was always sticking together at
          dances and stuff like that. And I was out there one Sunday, and
          Tony was over there because he used to come to visit the boys and
          maybe Marie, too. I don't know. But anyway, that's where I met
          Tony.[76]

Merle: Is that right?

Evelyn: Yeah. I didn't start going with him just right away then. But then
          he started walking me home from some of the dances.

---

[76]While there were several times in this conversation that we talked about how she first
met Dad, there is still more we would like to know. One of the things that was inadvertently not
discussed in this transcript was Dad's training as a barber. We do know that Dad earned a
diploma from the Sioux City Barber College on December 27, 1927. Dad talked about his
training at barber college in a letter dated October 12, 1927, from Dad to Grandpa and Grandma
Wilberding. For those of us who remember Dad's strictness and discipline, that letter from
Dad was an interesting surprise:

Dear Ma & Pa

Well I got back just fine, no trouble at all, took us about 2 ½ hours. He
played his banjo and sang, so it wasn't a bit lonesome; his folks moved and
live on 24th street now (there is an old red head living there with him, I
think, cause he talked about things 2 & 3 weeks ago, and about moving. I
think they have a high old time together. I would hate to be a mouse.)

I got my clippers. They sure burn off the hair. Boy, they work good and I
sure like this work. I've got 3rd chair today. We had demonstration last
nite, so I didn't write then. He showed us how to dye hair and eyebrows. I
trained a little last night, not very much. We . . . . I don't know any news at
all, so I will expect I may as well close for this time. Hoping you are all well
and waiting for a big long letter.

                                   I remain ever your loving son, Anthony

P.S. We boys sure have a big time together, always somebody joking
another, all are good clean sports. I got 3 bars of candy already, all for
nothing and a ten cent tip. Write soon. Good bye, dear folks.

Merle: Yeah. Where were the dances?

Evelyn: Well, the dances was in the Breda Hall what they got like . . . .

Merle: The old opera house?[77]

Evelyn: The opera house there, yeah.

Merle: Yeah, I know that.

Evelyn: Anyway, at one of those dances I danced with him.

Merle: Did he have a car then?

Evelyn: He didn't have one, well, I suppose he had one. It wasn't just too
          long when he got the new Ford, though.

Merle: The one you showed me the picture of?

Evelyn: Yeah, where I'm standing by . . . .

Merle: Yeah.

Evelyn: Yeah. That was in one of the earlier days.

---

[77]The Breda Opera House was a major source of amusement and sports activity during
the period from 1915 until 1940, hosting everything from boxing matches and basketball games
to movies and dances. And, in hindsight, no bigger dance may have been held than the dance
advertised in *The Breda News* as follows:

DANCE
BREDA OPERA HOUSE
TUESDAY, OCTOBER 25, 1932
MUSIC BY
LAWRENCE WELK and his ORCHESTRA
AMERICA'S BIGGEST LITTLE DANCE BAND

GENTS: $0.40   **LADIES $ 0.25**

Merle: Did you ever drive a car then?

Evelyn: I drove. I could drive our car at home.

Merle: Could you?

Evelyn: But we didn't have to have a license to drive, but Dad usually let us
          drive anyway, and some of the guys I'd go out with, they'd say, "Do
          you like to drive?" And I'd say, "Yeah." There was a few of them
          that, but . . . .

Merle: Were those open cars or were they closed?

Evelyn: No, they had like doors on the sides.

Merle: Did they?

Evelyn: Yeah. Any one that I was in. But I never was too good at it and I
          never, you didn't have to have a license for quite a while, you know.

Merle: So you never had a license?

Evelyn: I never had a license.

Merle: Didn't you take a ride in an airplane when you were young?

Evelyn: Yeah, it was a small plane with an open back seat.

Merle: . . . . with an open cockpit? Was that before you were married?

Evelyn: Oh, yeah. I was going with Tony at the time but my brother, Frank,
          he was just nuts about airplanes. He always wanted to be an
          aviator. And, so they had a show out there and they were taking up

passengers. It was an open cockpit, you know. And they could take . . . .

Merle: Where was this at, in Carroll?

Evelyn: At Carroll. It was right east of Carroll where they had some kind of a place where that plane was and they was taking people up for a ride. You'd have to pay so much for a ride to get up and they'd go over the town.

Merle: This was an open cockpit?

Evelyn: Yeah. Windy.

Merle: There were two people, one pilot and you?

Evelyn: One pilot and two people.

Merle: And you'd sit in the back behind the pilot?

Evelyn: Yeah, yeah. And I wanted to go up so bad.

Merle: Did you really?

Evelyn: Yeah. Tony wouldn't go. You wouldn't get his foot off the floor. So my brother Frank was there and he loved them, and Frank says I'll go along with you up there. He had had a ride already. So Frank and I went up there. But Tony wouldn't go.

Merle: No kidding. You're probably the only one who's been in the open cockpit in a plane. I don't know anybody else. Did you have goggles on? Did they give you goggles?

Evelyn: No, no. You just let your hair blow and it was, you couldn't hardly talk in there. It was windy. But he flew over the town and stuff and got down safe and my hair was just all tangled up but Tony wouldn't go. Even with me there.[78]

Merle: When you got married, you were living in that house then with the stone baskets there in town?

Evelyn: Right.

Merle: How long did Grandpa Goecke live there then?

Evelyn: He lived in that house in Breda until after Florence got married. Let's see now, that wasn't more than two or three years before he moved again – I think that's how it was.

Merle: Then is that when he moved up to Spirit Lake?[79]

Evelyn: Yeah.

Merle: Is that right? So he was up there for a long, long time.

---

[78]It is difficult to visualize Mom taking a ride in an open cockpit airplane. In my 40 years of knowing Mom, I can only remember Mom taking two airplane trips: when she and Dad flew to Ft. Lauderdale, Florida, in 1965 to visit Larry and Diane, and when she and my sister, Sal, flew to Dayton, Ohio, in 1981 to visit me, and my children, Abigail and James.

[79]For many years, Grandpa William Goecke had his "High View" cottage on Spirit Lake, Iowa. It was a very small cottage with little heat other than a small oil burning heater used for cool evenings in the spring and fall. (During the winter, Grandpa Goecke typically lived in a rooming house in Milford.) There were only a couple of small rooms in the cottage and one of those rooms had all of its walls covered with checkerboards, one of his favorite hobbies. Grandpa Goecke fished almost every day, using a small boat outfitted with a small outboard motor. (In 1996, when I built a home on Lake Willoughby in Vermont, I named it "High View" because it sits atop a bluff overlooking the lake and because I wanted to commemorate Grandpa Goecke's cottage on Spirit Lake, Iowa. Spirit Lake is well known in Iowa not only for its beautiful lake and surrounding country side, but also for the "Spirit Lake Massacre" in 1857 by Inkpadutah and his band of Sioux Indians which is detailed in *Commemorative Book of Spirit Lake Centennial* (1979) and MacKinlay Kantor's book, *Spirit Lake* (1955).

Clarence Goecke, Mamie Goecke Schelle,
Evelyn Goecke Wilberding, William Goecke, Teresa Goecke Oswald,
Florence Goecke Wolterman, Lizzie Goecke Wernimont

Evelyn: Yeah, yeah, he was. Let's see how old would he have been when I got married? Have you got it on there when he was . . . .

Merle: Well, I do some place. He got married in 1896. Oh, let's see. I don't have that right here.

Evelyn: I'll look in the Bible then. He was born in 1830. Oh, no. That's my Grandpa. Frank Goecke, that Grandpa. It's William Goecke, yeah. He was born in 1873. And he died in 1963. But, anyway, after I got married, Lawrence kept coming up to see Florence and then she was married in about two years.

Merle: Did you say before that Grandpa Goecke built that house with his brother, John?

Evelyn: Yeah.

92

Merle: That was his brother, John. Okay. Which, did he have a brother, Joe, too?

Evelyn: Yeah, he had, Joe was on the farm, yeah.

Merle: Okay.

Evelyn: Joe lived across the road from us and that place is sold now.

Merle: Which one is the one that had the Sac City farm?

Evelyn: Oh, that was Joe's son.

Merle: That was Joe's son.

Evelyn: Yeah, yeah.

Merle: And was Art Goecke a son of Joe also?

Evelyn: Yes, uh huh. Let's see. They had Hilda, Addie, and Edward they had, and Art Goecke and Joe Goecke and Sealy and Florentine. They had.

Merle: Which was the one from Sac City? Which Goeckes were they?

Evelyn: That is, let's see. Did you write the names down?

Merle: Art, Florentine, Edward, Hilda – was there another Joe? Did Joe have a Joe?

Evelyn: Yeah. Yeah, this was Joe up there at Sac City, yeah. That's right.

Merle: Joe's son lives ten miles away from me in Middletown, Ohio.

Evelyn: Joe's son does?

Merle: Yeah.

Evelyn: You're kidding.

Merle: I had some dealings, he was the director of sanitation for the City of
          Middletown and I had to go in to deal with him and, you know, I
          saw his name was Goecke and then he told me he went to the
          University of Iowa, and after that he was telling me that he had
          grown up in Sac City and he remembered coming to our farm with
          Ronnie Goecke when he was real young. He's about Larry's age.

Evelyn: Well, I bet he was home then. See the Joe Goeckes celebrated their
          golden wedding.[80]

Merle: Is that right?

Evelyn: Yeah, and Mamie and Charlie and Clarence and Clara and I went up
          there. They were having open house up at Sac City. And they had
          some darn good looking kids there.

Merle: So Grandpa moved up to Spirit Lake in what, before 1940 then?
          1935? 1940?

Evelyn: Well, let's see. When would Florence have got married? '29? '30?
          '32?

Merle: You got married in '29. So.

------

[80]The golden wedding was being celebrated by Joe Goecke, Jr., a nephew of Grandpa
William Goecke. Joe Goecke, Jr.'s brother was Art Goecke whose children Ronnie, Janet, and
Myrna attended school in Breda for the first few years before they moved to Mt. Carmel.

94

Evelyn: He moved up there. He was up there for quite a few years.[81] First
          he stayed out there on that east or went out on that east place or
          something. I don't know. I thought he, I can't even remember what
          he done after Florence got married.

Merle: Is Florence on that picture there?[82]

Evelyn: No, no.

Merle: No, was that before she was born?

Evelyn: Yeah, that was before she was born.

Merle: Is that right?

Evelyn: Uh huh. Clarence is the baby in . . . .

Merle: Yeah, yeah, I saw the names on the back and it didn't hit me right
          away but that picture . . . .

Evelyn: Yeah, Florence was I said about seven years younger than me.

Merle: She is?

Evelyn: Yeah. That's why I never thought when Lawrence would come
          along, you know, with Tony through the winter time, you know,

---

[81]The obituary for Grandpa William Goecke stated that his wife (Grandma Margaret
Boecker Goecke) died in 1930 and that he moved to Spirit Lake in 1934 and stayed there until
1956.

[82]In the Goecke family picture to which Mom is referring (and which is included in this
book), it is striking to note the physical resemblance between Grandpa William Goecke and
brother Ed Wilberding, and the physical resemblance between Mom (sitting on Grandpa
Goecke's lap) and many of Mom's grand children and great-grand children.

The William Goecke Family
Back row: Teresa, Frank, & Lizzie
Middle row: William Goecke & Margaret Boecker Goecke
Front row: Evelyn, Mamie, & Clarence

and he'd sit there and we'd all talk and play cards or something. And then Florence got married and Dad was still living in town there.

Merle: Where was your wedding when you got married? At that house?

Evelyn: Yeah. I didn't have a big wedding like most of the kids.

Merle: Uh huh.

Evelyn: But it was in the winter time. We was lucky to even have . . . .

Merle: It was cold then.

Evelyn: We had all we could do to get home. We went to Carroll to have our pictures taken, and it snowed so hard on the way home we were just glad to get home. Grandma and Grandpa left pretty quick after we had dinner at our house. And then they left pretty quick to get back home on the farm, you know, before they'd get snowed in there, too. And we wanted to go to Sioux City on a wedding trip. But we never got out. It was just the worse snow storm of the year. It was almost like when Ed was born.

Merle: You just went to Carroll for the pictures and came home and stayed?

Evelyn: Stayed at home.

Merle: Happy to be home from the way it sounds.

Evelyn: Happy to be home, yeah. Yeah, and just a few of the relatives stuck around there. They didn't amount to much.

Merle: Uh huh. But the wedding was in the Breda church.[83]

Evelyn: Yeah, oh, yeah.

Merle: Did you have booze at it?

Evelyn: No, no. My Dad wouldn't have no booze in the house.

Merle: Is that right? Your Dad never drank a drop.

Evelyn: My Dad never drank. That's why we're all non-drinkers.

Merle: Yeah. Nobody's ever, well, Dad never drank either.

---

[83]The write-up in *The Breda News* for Mom's wedding was in part as follows:

Miss Evelyn Goecke and Anthony Wilberding Married at St Bernard's

A pretty wedding took place at St. Bernard's Church Tuesday morning, January 22nd, when Miss Evelyn Louisa Goecke became the bride of Mr. Anthony F. Wilberding. Rev. J.A. Schulte performed the ceremony during a nuptial mass which united the young couple in the holy bands of wedlock, and was witnessed by a large number of relatives and friends.

The bride looked beautiful in her wedding gown of white flat crepe trimmed with silver lace, with bridal veil arranged with lace and rhinestones and wore silver slippers. She carried a bouquet of carnations and lilies of the valley. The bridesmaid, Miss Florence Goecke, sister of the bride, wore a gown of coral georgette over satin with hat and slippers to match and carried a bouquet of pink carnations. The bridegroom was attended by Rudolph Wessling.

The bride is a daughter of Mr. and Mrs. Wm. Goecke and has lived in this locality all her life. She is a pleasant and accomplished young lady and is held in high esteem by her large circle of friends. The bridegroom is a son of Mr. and Mrs. Frank Wilberding. He has resided in this vicinity all his life and is an industrious young farmer. The young couple left on a short honeymoon trip after which they will make their home on the groom's father's farm southwest of Breda.

98

Evelyn Wilberding & Anthony Wilberding
January 22, 1929

Evelyn: No. Tony never, no his folks weren't drinkers either. I mean you can drink a little bit once in a while but we never had drinks at our house. We'd keep like that, Grandpa Wilberding especially always liked that.

Merle: Brandy.

Evelyn: Brandy.

Merle: Yeah, apricot brandy. I remember him drinking the apricot brandy or at least had it in his house.

Evelyn: Yeah, he always had it and I tried to keep some in the house, too, sometimes when I . . . and if you put it in hot water you'll go to sleep.

Merle: Who was that Rudy Wessling? How was he related?

Evelyn: Well, he was the, you know, like a second cousin of Tony's.[84]

Merle: What's he to Gertrude? He's like a first cousin to Gertrude?

Evelyn: Well, it would be about the same with Gertrude as, their parents were cousins.

Merle: Just the same. Uh huh, uh huh. Was he a close friend of Dad's? How did he get to be the best man?

Evelyn: Tony was very short on finding any one to stand with him because there wasn't any relation really.

------

[84]As always, Mom had this relationship correctly identified. They were second cousins. Rudy Wessling was the son of Aloysius "Louis" Wessling and Mary Welte. He was born on July 3, 1907 – just a few months after Dad was born. Rudy's grandparents were Joseph Wessling and Christina Hoelter. This Joseph Wessling was a brother to Barney Wessling, the father of Grandma Mary (Wessling) Wilberding.

Merle: Uh huh, uh huh.

Evelyn: He didn't have any close relation and we used to kind of visit around
          and stuff with Rudy. In the first place Rudy was at our wedding.
          He was going with my cousin at the time.

Merle: Is that right?

Evelyn: Yeah.

Merle: Okay, so when you got married, you went to Carroll and got the
          pictures taken and came home, you went home to that wash house
          on the east place? Is that where you went the first night?

Evelyn: Oh, no. I mean we came back to Breda.

Merle: Okay.

Evelyn: When we had our pictures taken the day of the wedding.

Merle: Yeah.

Evelyn: And we went there right after dinner so we'd get back and that's all
          we had to do. You couldn't hardly see ahead of you it was storming
          so bad.

Merle: Okay, and then you came back to Breda for the reception?

Evelyn: Yeah.

Merle: Or to get together anyway.

Evelyn: Yeah, to get together and it was just family.

Merle: Yeah.

Evelyn: Tony didn't have much family and there was just my brothers and sister really that was there, and the neighbor ladies did the cooking and served the lunch and stuff like that.

Merle: Is that right. Did you have a cake?

Evelyn: I had a cake. Just like a . . . not a fancy decorated one. It probably was an anglefood cake about that high.

Merle: Where did you get the dress?

Evelyn: I made it.

Merle: You made it?

Evelyn: I made it.[85]

Merle: Do you still have the dress?

Evelyn: I don't think so. I think that stayed out in that one trunk and I never moved it to town, and I had a fur coat that stayed out there, too.

Merle: Really? And that's the trunk that's not there anymore?

Evelyn: Uh huh. And I had silver slippers and I made Florence's bridesmaid dress.

Merle: What did you do, buy the thing, the fabric?

Evelyn: Fabric, yeah.

Merle: At the . . . . Is Mrs. Wilson still running it?

---

[85]The elegance of the wedding dress that Mom made for herself reflects her skills as a seamstress – skills she repeated time and time again over the years as she made skirts, blouses, and dresses for the girls and shirts for the boys.

Evelyn: Oh, yeah.

Merle: Is that where you went up and got it?

Evelyn: Probably so.

Merle: So you made the dress, huh?

Evelyn: Yeah, I made the dress.

Merle: Did you have a pattern or did you need to get patterns from the store?

Evelyn: Oh, yeah, we always we made most of our clothes those days. In fact, I never had a ready-made dress to wear until then. Clarence was married just the week before.

Merle: Were they?

Evelyn: Yeah, just the week before. And I didn't know what to wear there, and I had this dress made and I planned that for my second wedding dress. And then I didn't know what to wear to Clara's so then I went to Carroll with somebody and I bought a dress, and that was the first boughten dress that I had.

Merle: Is that right, the one you wore to Clarence's wedding?

Evelyn: No, I had made the one to Clarence's wedding and I bought one to have a second dress for the wedding.

Merle: Oh, okay.

Evelyn: Yeah. I don't know if I put that on that night or not or if I kept the veil on. But anyway, oh, yeah, I did have it on there because we have some pictures taken. There's one in that old book of me with kind of little ruffles on the bottom and then there's a priest, Father Schulte was it?

Merle: Yeah, Father Schulte was the priest.

Evelyn: Yeah, I think we took that later on in the afternoon. I think I took the veil off and put that other dress on or something that I had that picture taken in it.

Merle: Did Grandpa Goecke ever talk about or did you know where he got married when it was at Sugar Creek? Where is that?[86]

Evelyn: Well, it's out in the eastern part of Iowa, and Mamie said she's been around there where they'd go to Dubuque and stuff, and they said the town is just gone.

Merle: Right around Dubuque and the town.

Evelyn: Well, I think it's lower down on the map than that. You don't find it on any map or anything, but it was in Iowa, and let's see, where was Mamie baptized? She was born out there.

---

[86]The community known as Sugar Creek is located in Waterford Township, Clinton County, Iowa. The Catholic Church was officially known as St Joseph's of Brown's because it was located at Brown's Station, a railroad depot then operated by the Milwaukee Railroad. The Church is three and one-half miles north of Charlotte, Iowa, just south of Preston, Iowa. The Church was built in 1896 and its official address is Preston, Iowa. The official county records of the marriage between William Goecke and Margaret Boecker are found in Volume 2, Page 160, of the Marriage Records of Clinton County, Iowa. It certifies that on October 20, 1896, Reverend W.B. Sassen married "Wm Gerk" and "Maggie Bocker" at Sugar Creek, Iowa. It also lists William Goecke's birthplace as "Mt. Carroll" which I thought was a simple confusion between his actual birthplace of Carroll, Iowa, and the fact that he was living in the Mt. Carmel community in Carroll County. But, when I was looking at the map to confirm that Sugar Creek was just west of the Mississippi River, I noticed that there is a "Mt. Carroll, Illinois" just east of the Mississippi River and the name of that community may have ontributed to the misidentification of his birthplace in the official records.

Merle: Okay, but you were born around here, around Breda.

Evelyn: Yeah, I was born in Carroll, 'cause I went in once to look.

Merle: Were you born in a hospital?

Evelyn: Oh, no. Nobody went to the hospital.

Merle: But all your, Lois and Alice and all those, were born in the hospital
           weren't they?

Evelyn: Yeah, they was, yeah. But like Mamie, the only one Mamie had in
           the hospital was Ron.[87]

Merle: Is that right?

Evelyn: Yeah, I was out there when Merlin was born.[88] They used to have
           what they would call mid-wives. They'd go out there and they'd go
           take care of her and they'd get the doctor out there and . . . .

---

[87]Mom is referring to Ron Schelle who was born on May 23, 1943. Ron was the son
of Mamie (Mom's sister) and Charlie Schelle.

[88]Mom is referring to Merlin Schelle, another son of Mamie and Charlie Schelle.
Merlin was in the Army during World War II. While he was overseas, he wrote several letters
to Mom and Dad, both of which were written shortly after I was born on March 8, 1944. In a
letter dated March 30, 1944, from "somewhere in England," he asked "How is the new boy
getting along by now?" Later, on July 25, 1944, from "somewhere in France," he wrote that he
guessed that Mom and Dad were "busy with the harvest. I know that last year about this time
I was working for you, wasn't I. I'd sure like to be back there again." What is interesting about
both of these letters is that they seem to be old-fashioned copies, reduced in size, and both have
the Censor's Stamp as being approved by the Army Base Examiner. Shortly after this time, and
shortly after the Battle of the Bulge, the German Army broke through the Allied lines on
January 12, 1945, and captured a number of soldiers, including Merlin Schelle. He served as
a prisoner of war at Heppenheim, Germany, and was liberated on March 28, 1945 (*Breda
Centennial Book*, Pages 335-336).

Merle: What happened when the banks failed? Didn't the Breda Bank fail, stop, or close, or what happened?

Evelyn: Oh, yeah. Well, you just didn't buy anything.

Merle: Did Dad have money in the banks then?

Evelyn: I don't think he had too much in there. Grandpa had quite a bit in there.

Merle: Did he lose it forever?

Evelyn: He got some of it back but not all of it.

Merle: Really?

Evelyn: Yeah. But we could live, see that was one of the first years that we was married. We didn't have much of a family and, as I said, we always butchered your own meat and had always your potatoes and you'd can stuff in the summer time and had eggs and milk.[89]

Merle: So in the Depression you got by because you could make your own food.

Evelyn: Yeah, we could make our own food, we never suffered a bit from that.

Merle: And the farm. Was it paid for then at the start?

Evelyn: Oh, yeah. Grandpa had them both.

---

[89]When Mom remembers about how she used to "can stuff" in the summer, she summed up an enormous project in two little words. As I noted earlier, the fruit room in the basement on the West Place was big and generally well stocked. But this was because we had spent many long days harvesting fruits and vegetables from the huge garden and grove of fruit trees and then canning these fruits and vegetables into the Mason jars with the sealed lids.

Merle: Grandpa never borrowed to buy the farms did he? He just paid for them.[90]

Evelyn: That I don't know, what he done when he bought them. But they were all paid for when we got there, and then he signed the one over to me and Tony after we got married. If I would have been smart, I never would have signed that thing.[91]

Merle: Grandpa was pretty good-hearted.

Evelyn: Oh, he was. Without him I don't know how we would have got along during that Depression and stuff, because even the food and stuff, you know, they had all those potatoes that they would put in their basement. We had planted some, but then in the canned goods and eggs and milk and cream and stuff, there wasn't much . . . .

Merle: You really didn't buy much of it did you?

Evelyn: Uh uh.

Merle: You made your, I suppose you had to buy some ingredients to make bread, like yeast and I suppose flour.

Evelyn: Yeah, flour and yeast, but outside of that, you could live very good on what we had. We always made our own cakes and pies and

---

[90]As I noted in Footnote # 1, Anton had obtained a mortgage from Connecticut General Life Insurance Company to acquire the East Place in 1894. However, there is no court record suggesting that Grandpa Frank Wilberding had to borrow any money to buy the East Place from the Estate of Anton Wilberding or to buy the West Place from his in-laws.

[91]Again, this is a good example of Mom's self deprecating comments. We all remember, of course, that Dad always did 100% of the taxes and money management. Yet, after Dad died, we all recognized, acknowledged, and applauded Mom as she very quickly demonstrated that she had great acumen in the family's financial affairs. I still remember sitting down with her while she explained how she had "laddered" all of her certificates of deposit to maximize her return and minimize her risk.

didn't buy stuff from the stores nor can goods. You'd can it
yourself, . . . what you'd get out of the garden.

Merle: They didn't have cake mixes. You just made everything from . . . .

Evelyn: That's right. You made everything from scratch. Pies and stuff, that
was just lard and flour . . . .

END OF TAPE

* * * * * * * ** * * * * * *

[At this point there was a knock on the kitchen door and a neighbor came in
to visit with Mom and me, and we stopped taping the conversation.]

EPILOGUE

Shortly after this conversation, Mom was hospitalized at St. Anthony's Hospital in Carroll, Iowa, and then, in January 1985, Mom went into St. Anthony's Nursing Home. Her condition stabilized for the next few months before it worsened in late May. Mom died on June 14, 1985, on her favorite feast day, the Feast of the Sacred Heart. The Mass of the Christian Burial was celebrated on June 17, 1985, at St. Bernard's Church in Breda, Iowa, and she was then laid to rest at the Breda cemetery. Mom was, for all of us, a gift from God, and we gave her back to God with gratitude, pride, and love, to rest in peace.

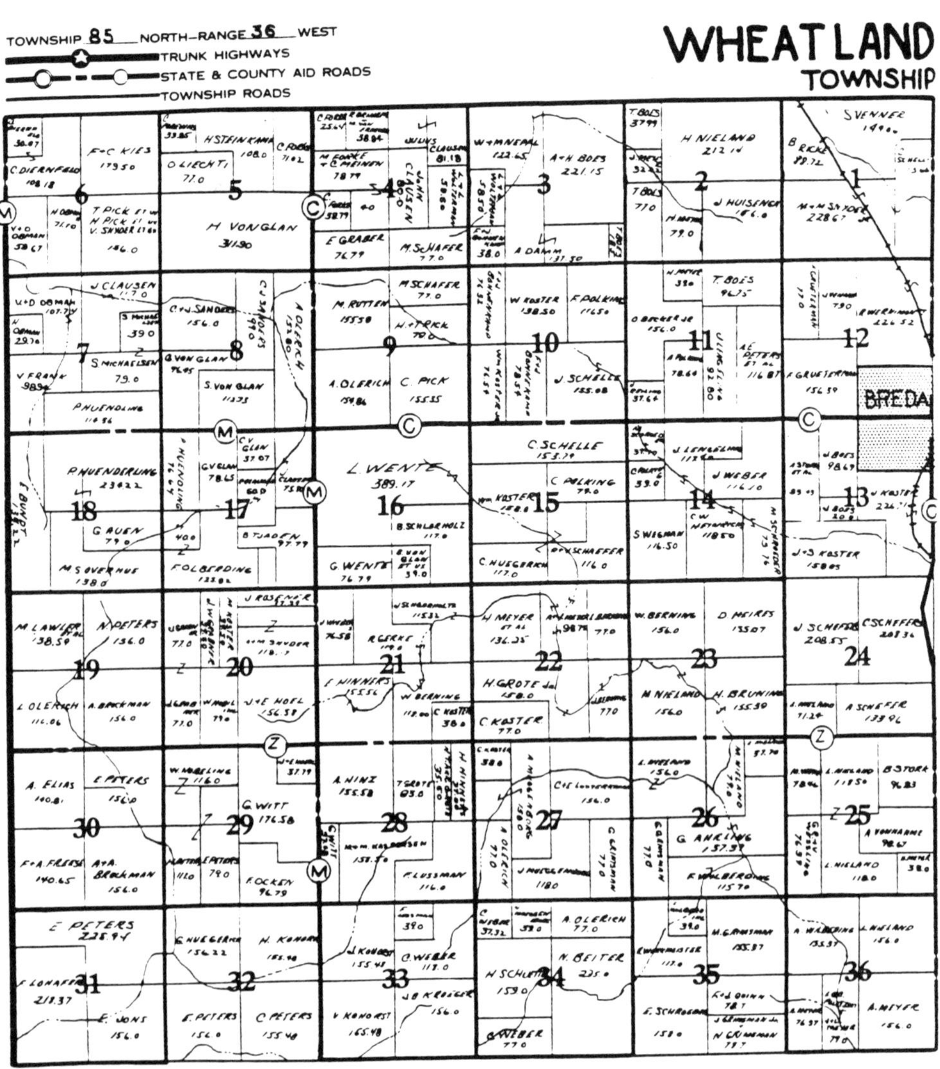

On this 1950 Plat Map of Wheatland Township, in Carroll County, Iowa, the East Place is in the northwest quarter of Section 36. This parcel is identified by "A. Wilberding" and shaded yellow. One hundred twenty acres of the West Place are in the south half of the southeast quarter of Section 26, and the other forty acres of the West Place are in the northwest quarter of Section 35. These parcels are identified by "F Wilberding" and shaded green.

WILBERDING FAMILY TREE

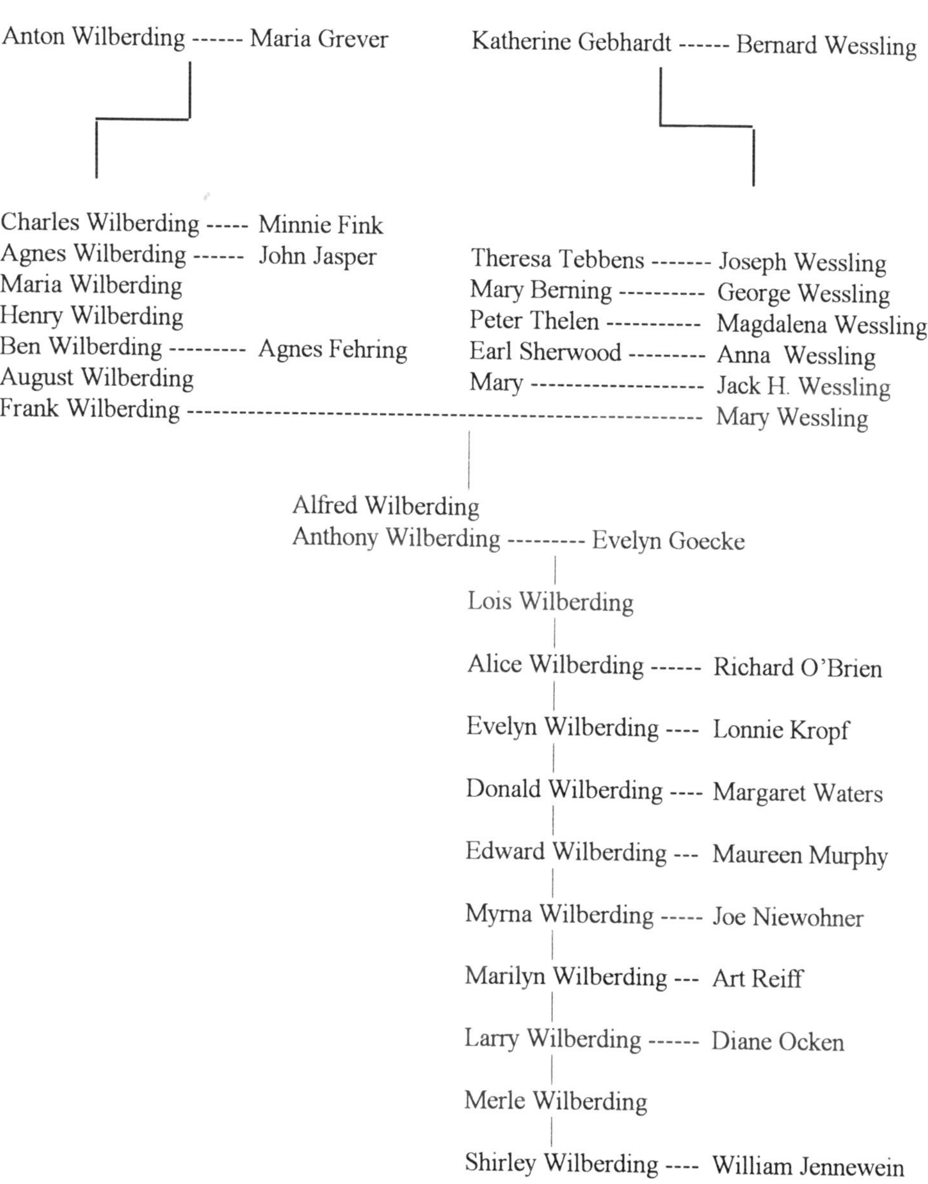

GOECKE   FAMILY TREE

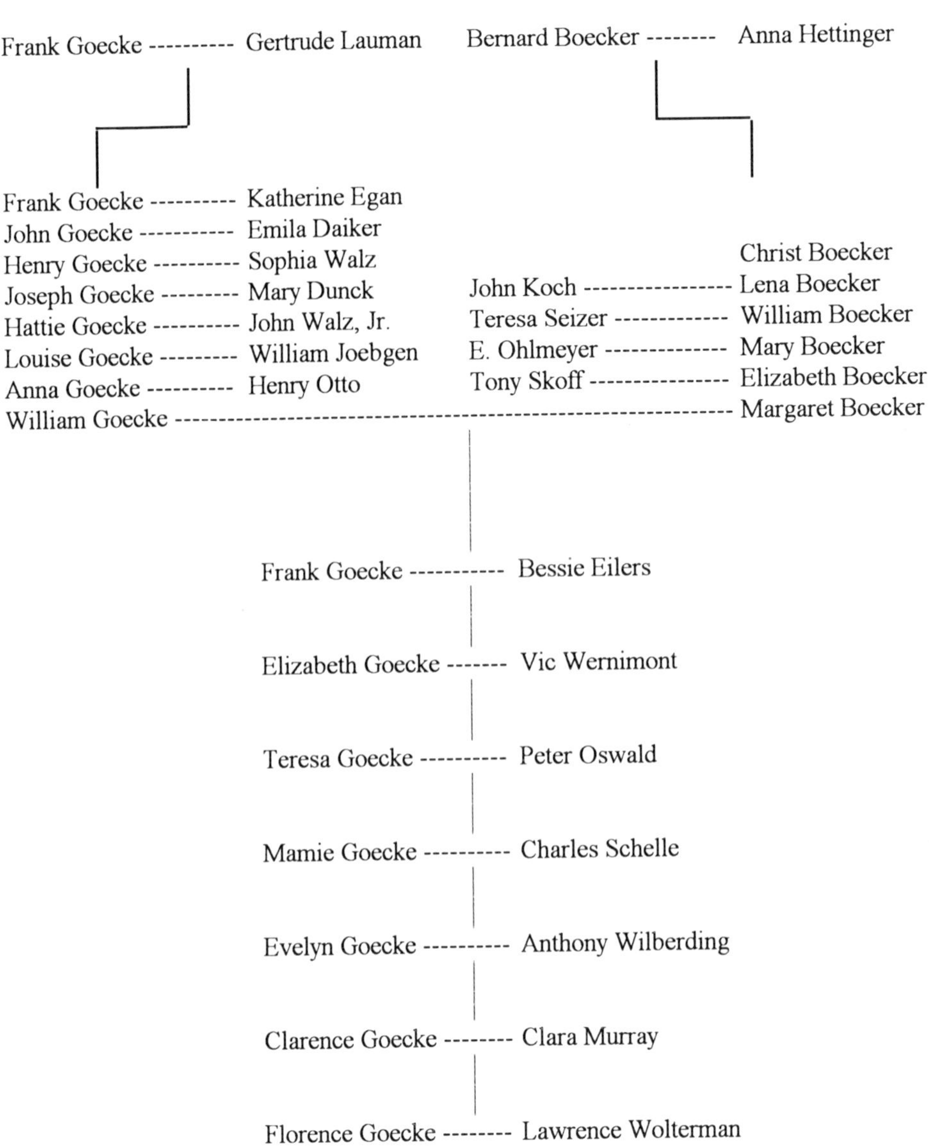

WESSLING FAMILY TREE

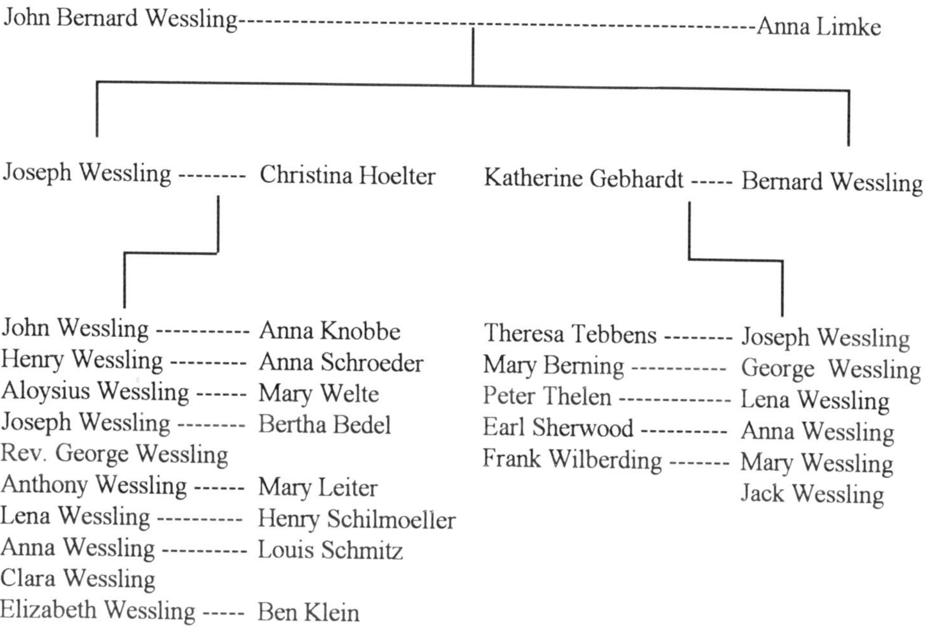

SCRAPBOOK POETRY

In gathering the materials for this book, I was struck by some items in Mom's own scrapbook, in which she had pasted some wild flowers, some birth announcements, and the clippings recording the death of her own mother, Margaret Boecker Goecke, who died on February 4, 1930, only days after death of Mom's first child, Lois, who died in infancy (January 11-13, 1930). Along with these death notices, there were several pieces of poetry which Mom had clipped out of a newspaper and pasted in her scrapbook. Obviously, these poems were helpful to Mom in remembering her own mother, and these poems will also enrich our understanding of Mom. Those poems appear on the following pages, after which there are blank Scrapbook Pages which can be used for individual clippings or other remembrances of Mom.

MEMORIES

"The trees are all in bloom"
My mother wrote to me:
"And fragrance comes into my room
From your favorite cherry tree."

Across the plains, she waits for me:
My mother – all alone –
She often writes and tells me
That she wishes for me home.

Then I do not see the fields
With harvest rich and bare
I can only see my mother waiting,
She is waiting for me there.

Sometimes I picture her at home
Amid the flowers and trees,
And feel, she sighs and dreams with me
Of poignant memories.

It is then I see the little home
With blooming shrub and tree
Oh, Mother, you alone can know
How memories comfort me.

Author Unknown

MOTHER

Her eyes are as blue as heaven's own sky
Her hair tinted silvery grey
Her face is a story of unselfish love,
Her smile is a sunshine ray.

Her voice is a song of devotion true
Lifted oft to console and in prayer.
And her hands are employed every golden hour
In patient and loving care.

Oh, Mother of mine, your dear face is impressed
Deep in my heart's best love:
You are always near in temptation and fear,
Guiding me to that realm up above.

And whenever the pleasure and transient joy
Of this world would lead me astray.
I can see your dear face, oh Mother of mine,
And contritely and humbly I pray.

Oh, my dear loving guide, when I think of you
I firmly resolve to trod
On the path of true glory and happiness fair,
On the path that leads to my God.

Gertrude Schuster

UNTITLED

One by one thy duties wait thee,
Let thy whole strength go to each,
Let no future dreams elate thee,
Learn thou first what these can teach.

One by one (bright gifts from heaven).
Joys are sent thee here below:
Take them readily when given,
Ready, too, to let them go.

A. A. Proctor

PRAYER OF A FARMER'S WIFE

Lord, help me find the simple beauties
That are lying in my way.
And to find some satisfaction,
E'en though small, in every day.

Make me see beauty everywhere,
This is my prayer, dear Lord, to you.
E'en if 'tis only in the stirring
Of a yellow cake in crock of blue.

Make me always be contented
With this simple, homely life.
And let me say, "I thank you, Lord,
For making me a farmer's wife."

Carmine Gray

MY SORROW

I knelt in sorrow by her side.
My beloved was dead;
Her lips no words they spake,
Her eyes no tears of sorrow shed.

My little home in Eldarado
Was hallowed as a shrine;
But shadows of the tomb
Have made is blest devine.

An old man shook with heavy sigh;
That quiet stillness there
Speaks the anguish and grief
No words can e'er declare.

I knelt in sorrow by her side,
Oh, few the words I said;
The house was oh so desolate,
For my mother dear was dead.

Prince Kerma

A MOTHER PRAYS

Father, be Thou close to me
When my baby stops here play,
Nestles close against my knee,
Gravest things to ask and say.

May I be her faithful friend
Swift to shed my household cares,
Always having time to spend
With her play and with her prayers.

For parted lips and trusting eyes.
Grant me wisdom, Lord, from you.
May my judgment, Lord, be wise;
May my answers all be true.

Catherine C Coblentz

SCRAPBOOK PAGES

SCRAPBOOK PAGES

SCRAPBOOK PAGES

SCRAPBOOK PAGES

SCRAPBOOK PAGES

SCRAPBOOK PAGES

SCRAPBOOK PAGES

SCRAPBOOK PAGES

SCRAPBOOK PAGES

SCRAPBOOK PAGES